Brittle Fracture and Damage of Brittle Materials and Composites

For Florence, my wife
For Natacha and Tatiana, my daughters
For my family

Series Editor
Jacques Lamon

Brittle Fracture and Damage of Brittle Materials and Composites

Statistical-Probabilistic Approaches

Jacques Lamon

First published 2016 in Great Britain and the United States by ISTE Press Ltd and Elsevier Ltd

ISTE Press Ltd
27-37 St George's Road
London SW19 4EU
UK

www.iste.co.uk

Elsevier Ltd
The Boulevard, Langford Lane
Kidlington, Oxford, OX5 1GB
UK

www.elsevier.com

British Library Cataloguing-in-Publication Data
A CIP record for this book is available from the British Library
Library of Congress Cataloging in Publication Data
A catalog record for this book is available from the Library of Congress
ISBN 978-1-78548-121-5

Printed and bound in the UK and US

Contents

Introduction

The fracture resistance of materials is an important issue for the reliability of systems. A lack of reliability wastes money and has slowed technological progress in many areas. Flaws are the principal source of fracture in many materials, whether brittle or ductile, whether nearly homogeneous or composite. They are introduced either during fabrication, or surface preparation, or during exposure to aggressive environments (e.g. oxidation and shock). The critical flaws act as stress concentrators and initiate cracks that propagate instantaneously to failure in the absence of crack arrest phenomena as encountered in brittle materials. In those brittle materials susceptible of crack arrest (such as continuous fiber reinforced ceramics), flaws initiate crack-induced damage. Fractures cannot be understood without a full description of microstructural features.

There are numerous brittle materials, including ceramics, glass, concrete, metals and polymers. Glass and ceramic fibers can be added to the list. Many industrial systems or structures use brittle materials. Furthermore, technological progress and the need for systems with high performances favors the development of new structural materials, which display interesting advantages such as a low density, a high resistance to elevated temperatures and aggressive environments, and good mechanical properties. These materials are the subject of important research and development programs.

It is still believed that fractures occur when nominal stress reaches a critical value, termed the strength of material. This stress is improperly considered to be a material property. It is often used as a fracture criterion. Griffith explained the failure of brittle materials. He showed on glass specimens that the low fracture strength observed in experiments was due to the presence of microscopic flaws. From this, fracture mechanics was developed. Fracture mechanics is the field of mechanics concerned with the study of the propagation of cracks in materials. Fracture mechanics concepts are used to predict the resistance to failure of materials and structures that contain a crack. Fracture toughness gives the fracture stress as a function of crack size. It is a common practice to assume that a flaw of some chosen size will be present in components, and use the linear fracture mechanics approach to calculate the stress required to propagate the crack. However, experimental failure data obtained for brittle materials are generally not reproducible, and they fluctuate by as much as an order of magnitude. The deterministic strength concept is questionable. Weibull argued that a statistical approach would be more appropriate. He proposed the so-called Weibull distribution, which he claimed applied to a wide range of problems including the strength of steel. The foundation of his theory has been questioned, and more physical statistical-probabilistic approaches to brittle fracture have thereafter been devised.

In brittle materials, flaws are numerous and randomly distributed. Their location in the material, their dimensions and their severity are generally unknown *a priori*. As a result, it is impossible to predict the behavior of brittle material under load with deterministic concepts. Fractures seem to be erratic. Brittle materials are often considered as unreliable.

The statistical approaches to brittle fractures provide functions that describe the distribution of fracture strengths. With the equations of probability, the occurrence of fracture can be predicted. There are two main types of probabilistic-statistical approaches to brittle fractures:

– the phenomenological and macroscopic approaches such as the Weibull model;

– the fundamental approaches that consider the flaws as physical entities. These approaches are based on flaw strength density functions. Flaw strength is defined using either the elemental strength concept or flaw size. These approaches make better predictions and are more robust than the Weibull approach because of the underlying flaw strength density function.

Fracture mechanics deals with the propagation of well-defined cracks exclusively, i.e. those cracks whose size and location are known. The probability of the presence of a flaw that initiates the crack must be introduced in the analysis. This is the added value of probabilistic approaches to brittle fracture. Similarly, the introduction of statistics in physics allowed for significant progress. Statistical physics uses methods of probability theory and statistics in solving physical problems. In particular, statistical mechanics provides a framework for relating the microscopic properties of individual atoms to the macroscopic properties of materials.

This book focuses on probabilistic approaches to brittle fracture with emphasis placed on those approaches which consider the flaws as physical entities, and more particularly the approach based on the multiaxial elemental strength concept for multiaxial problems. The following issues are discussed successively in the chapters of the book:

– the effects of flaw populations on fracture strength;

– the main statistical-probabilistic approaches to brittle fracture;

– the use of these methods for predictions of failure and effects induced by flaw populations (such as effects of size and loading mode);

– the application of these methods to component design;

– the methods of estimation of statistical parameters that define flaw strength distributions;

– the extension of these approaches to damage and failure of continuous fiber reinforced ceramic matrix composites.

Several examples of failure predictions are discussed that should be helpful to the readers. These case studies can be useful for a better understanding or for solving problems. Most examples come from high-performance ceramics, which represent quite well the family of brittle materials, and exemplify the type of fracture problems encountered with brittle materials.

This book will be of use to students, engineers and researchers who are interested in fracture and brittle materials. It tackles theoretical and experimental aspects. It proposes concepts, tools and methods to solve problems of fracture or design. Thus, for many brittle materials, their introduction in industrial systems requires appropriate approaches that are efficient to predict not only the conditions of fracture or damage in service, but also the associated risk of occurrence of these conditions.

1

Flaws in Materials

1.1. Introduction

Any material can contain flaws, i.e. materials are never free of flaws. According to dictionaries, a flaw is "an imperfection, often concealed, that impairs soundness". Here, a flaw is a heterogeneity that disrupts the theoretical order and introduces a discontinuity or a singularity. Flaws impede the working of materials and systems, as well as various optical, magnetic, mechanical, etc. properties. Various words are used for flaws, depending on length scale: dislocation, vacancy, impurity, interstitial (crystal defects), fault (geology), cavity, hole, etc. The flaws of interest in this book are those which are responsible for fractures. The occurrence of flaws is not completely avoidable in the processing, fabrication or service of a material/component. They may appear as cracks, voids, metallurgical inclusions, weld defects, design discontinuities or some combination thereof.

Flaws are the weakest link of materials. They may have any shape. In certain materials, their size can be as small as a few micrometers (engineering ceramics and glass) or a fraction of a micrometer (fibers of carbon, ceramic or glass). They act as stress concentrators: they cause stress to increase locally, so that the local stress can exceed the intrinsic strength. Intrinsic strength is the strength in the absence of flaws. In single crystals, it is the stress required to break atomic bonds. It is called the theoretical strength or the ideal cohesive strength. It depends essentially on atomic bonding force. In polycrystalline materials, it takes smaller value than the theoretical strength, because of the presence of grain boundaries.

The fracture stress of materials depends on characteristics of flaw populations. As a result, it depends on extrinsic factors that contribute to flaw criticality such as specimen size, loading conditions, etc. Its sensitivity to flaws is significant when flaw extension is instantaneous and causes catastrophic failure. Brittle materials are very sensitive to flaws. The class of brittle materials includes numerous materials such as ceramics, glass, concrete, metals and polymers (at temperature < 0.75 Tg (Tg = glass transition)). By contrast, in ductile materials (like metals) or in damageable materials (like continuous fiber reinforced composites), the effect of flaws on ultimate fracture is less critical because their propagation is hampered by crack arrest phenomena (such as dislocations, slip bands, debonding at interfaces, etc.). However, in ceramic matrix composites, flaws initiate damage.

Flaws must be regarded as fundamental constituents of materials. As a result of their presence, fracture strength exhibits several features that are discussed in this first chapter. These features need to be known and accounted for in order to make sound predictions of in-service failure of components. For convenience in the discussion, we will consider the case of engineering ceramics on which a large amount of research work has been produced. They are a representative class of brittle materials.

1.2. The theoretical strength and the intrinsic strength of materials

Flawless single crystals consist of arrays of atoms that form a regular lattice. Binding of atoms involves interatomic forces, resulting from the energy of crystal. A repulsive energy and a form of attractive energy contribute to total energy. The interatomic forces ensure cohesion. Fractures occur when the links between atoms in a plane are no longer realized, splitting the material apart. The elementary modes of fracture are:

– cleavage or opening mode: the fracture plane is perpendicular to stress direction (Figure 1.1);

– slipping or shearing mode: the fracture plane is parallel to the loading direction.

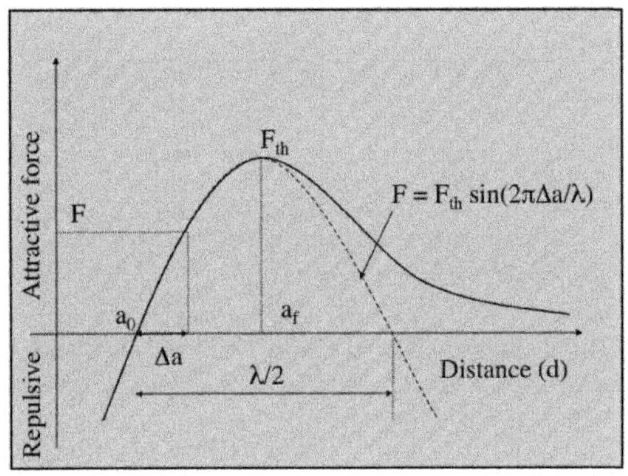

Figure 1.1. *Interatomic force-intensity curve derivative of the total energy: d is the distance between two atoms, a_0 is the distance for binding energy (0° K): $d = a_o + \Delta a$*

The opening mode governs brittle fracture. The theoretical strength (also termed ideal cohesive strength) is the stress required to break the interatomic links in the fracture plane. An exact calculation of the theoretical strength is possible for simple cases such as ionic crystals. Approximation considers the interatomic stress curve of a crystal replaced by a half sine wave (Figure 1.2).

$$\sigma_{th} = \sqrt{\frac{2E\gamma_s}{a_o}} \qquad [1.1]$$

E is the Young's modulus, γ_s is the specific fracture energy and a_0 is the interatomic distance.

Estimates of theoretical strength range between 0.05 and 0.1 E. The Young's modulus of crystalline materials with strong interatomic bonds (covalent or ionic) is several hundreds of GPa (Table 1.1). The theoretical strength is as large as 10 GPa.

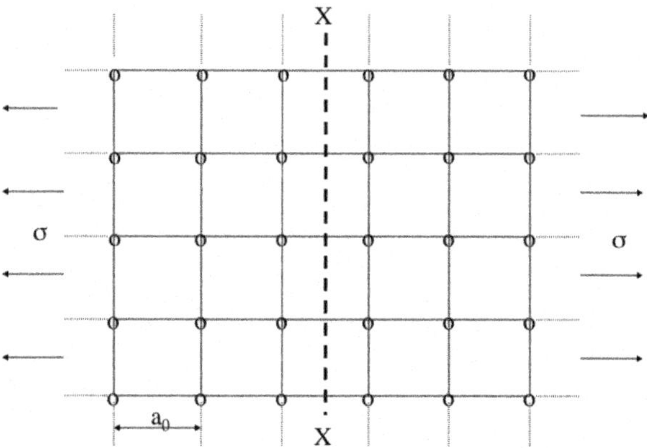

Figure 1.2. *Brittle fracture by cleavage: fracture of interatomic bonds across plane XX, perpendicular to the direction of tensile stresses (σ)*

Material	E (GPa)	σ_R (GPa)	σ_R /E
Aluminum	70	0.09 – 0.15	1.3 – 2.1 10^{-3}
Copper	112	0.12 – 0.40	1.1 – 3.5 10^{-3}
Mild steel	210	0.18 – 0.50	0.85 – 2.4 10^{-3}
High strength steel	210	1 – 3	4.7 – 14.2 10^{-3}
Glass	71	0.03 – 0.09	0.42 – 1.26 10^{-3}
Alumina	380	0.2 – 0.3	0.53 – 0.8 10^{-3}
Silicon Carbide	410	0.2 – 0.5	0.48 – 1.2 10^{-3}
Silicon Nitride	310	0.3 – 0.85	0.9 – 2.7 10^{-3}
Zirconia	200	0.2 – 0.5	1 – 2.5 10^{-3}
SiAlONs	300	0.5 – 0.83	1.67 – 2.7 10^{-3}
Carbon fibers	450	1.5 – 2.5	3.3 – 5.5 10^{-3}
SiC fibers	200 – 300	3.0	1 – 1.5 10^{-2}
EGlass fibers	72	1.5	2 10^{-2}

Table 1.1. *Young's moduli and fracture strengths of metals, glass and high-performance ceramics.*

1.3. The fracture strength of materials

The tensile strength (also called ultimate strength) measured on a variety of materials is generally less than 1 GPa (Table 1.1): it is less than 0.9 GPa for ceramics, and less than 3 GPa for single filaments. It may be more than 100 times as small as the theoretical strength.

This discrepancy was resolved by Griffith, who proposed that the theoretical strength required to break the interatomic bonds is reached at the root of a sharp and narrow crack. He showed that such a crack in a plate can become unstable at relatively low nominal stresses. This phenomenon can be explained by the stress-concentrating action of a crack or a flaw. It results from the presence of free edges (crack lips) that allow large deformations of the atomic layers at crack tip. As a result, higher stresses operate in this area. This mechanism is shown in Figure 1.3 using imaginary line forces. In a solid subject to uniaxial tension, the line forces are parallel. This initial order is disturbed when a crack is introduced. Line forces become longer at crack tip. If these lines correspond to chains of atoms, it comes that the interatomic links are more elongated at crack tip. Springs can also be used for the demonstration.

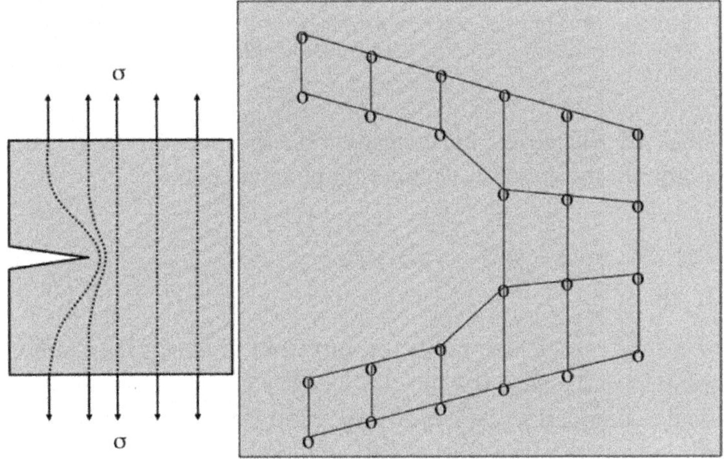

Figure 1.3. *Stress concentration at crack tip under tensile stresses perpendicular to crack plane: a) line forces, b) dilatation of interatomic lattice at crack tip*

1.3.1. *Influence of a crack and a flaw*

The local tensile stress (σ_T) at the root of a crack acting across the extension of the plane of the crack as shown in Figure 1.4 is related to the applied tensile stress (σ is related to the applied) [ING 13] formula:

$$\sigma_T = \sigma\left(1 + 2\sqrt{\frac{a}{\rho}}\right) \quad [1.2]$$

where 2a is the length and ρ is the radius of curvature at crack tip.

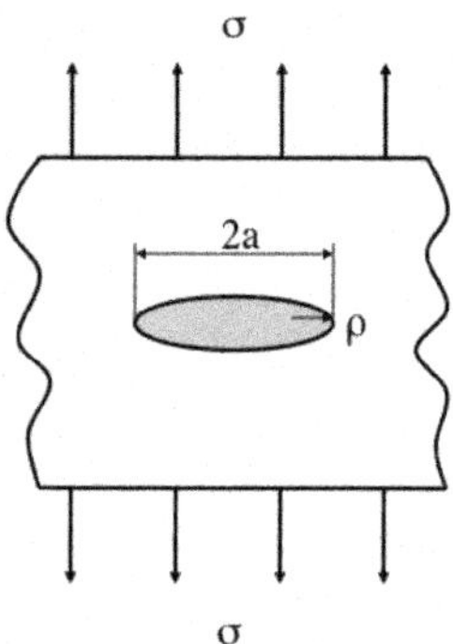

Figure 1.4. *Through crack of length 2a with its plane normal to the direction of a uniaxial tensile stress (σ) in an infinite plate*

Values of the stress concentration factor derived from formula [1.2] highlight the stress-concentrating phenomenon:

$$K_t = \frac{\sigma_T}{\sigma} = 1 + 2\sqrt{\frac{a}{\rho}} \quad [1.3]$$

For a sharp crack, the radius of curvature is generally taken to be the interatomic distance 0.2 nm. For crack size 2a = 2 μm or k_t= 143. This value compares well with the ratio of theoretical to tensile strength indicated above. For a stiff ceramic having Young's modulus = 500 GPa and theoretical strength = 50 GPa, the strength of 0.35 GPa calculated using (equation [1.3]) agrees well with the values reported in Table 1.1.

The stress concentration factor reflects flaw severity under uniaxial tension. It depends on flaw dimensions. At a fracture, the following formula for the stress concentration factor is derived from equation [1.3]:

$$\frac{\sigma_{th}}{\sigma_R} = 1 + 2\sqrt{\frac{a_c}{\rho_c}} \qquad [1.4]$$

where a_c and ρ_c are the dimensions of the flaw that caused fracture (the critical flaw). σ_{th} refers to the intrinsic strength, i.e. the material strength in the absence of flaws. It is probably smaller than the theoretical strength defined above as the ideal cohesion strength. σ_R is the corresponding applied tensile stress. It represents the tensile strength.

According to equation [1.3], flaw severity under uniaxial tension increases in inverse proportion of the radius of curvature at flaw tip. Sharp cracks are the most severe flaws. The stress concentration factor can be as large as 150, depending on crack length. By contrast, when the radius of curvature takes large values, the stress concentration factor decreases to 1, and the flaws become innocuous. For spherical voids ($a_c = \rho_c$), $K_t = 3$: the tensile strength is three times as small as the intrinsic strength. Compared to cracks, spherical voids are quite innocuous. Figure 1.5 shows the relationship between tensile strength and intrinsic strength for a variety of commercial glass. The tensile strengths are less than 0.5 GPa, whereas the intrinsic strengths exceed 3.5 GPa.

1.3.2. *The resistance to crack extension*

The local stress-state induced by a crack is approached by fracture mechanics concepts. The stress intensity factor is used to predict the stress-state near the tip of a crack caused by a remote load or residual stresses (Figure 1.6). Linear elastic theory predicts that the stress distribution near the crack tip has the form:

$$\sigma_{ij} = \frac{K_\alpha}{\sqrt{2\pi r}} f_{ij}^{\alpha}(\theta) + 0\sqrt{\frac{1}{r}} \qquad [1.5]$$

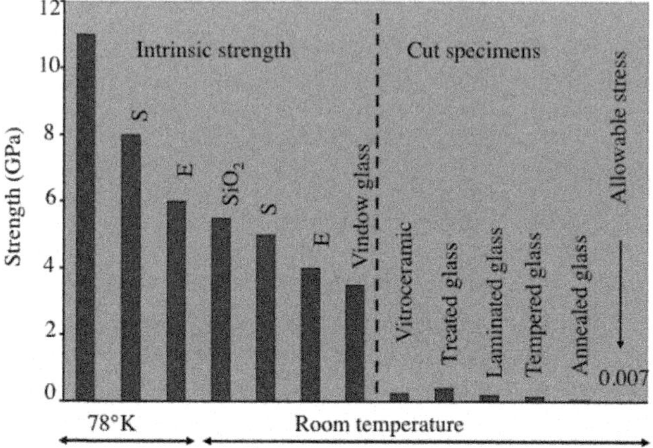

Figure 1.5. *Fracture strengths for various types of glass as functions of processing, temperature and origin*

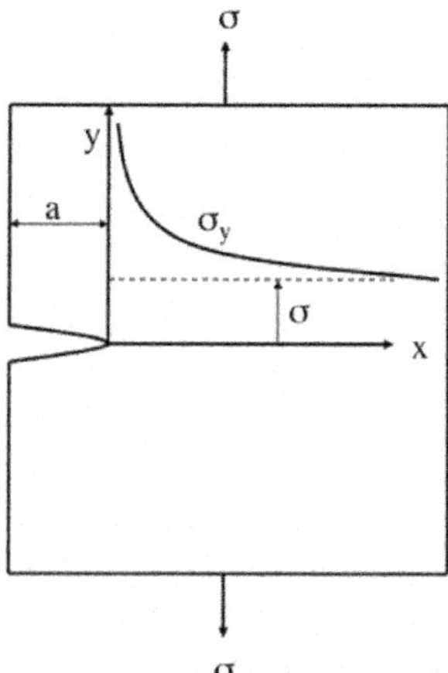

Figure 1.6. *Stress field near crack tip in a volume element subjected to a tensile stress. Only the opening stress component σ_y parallel to loading direction is shown*

K_α is the stress intensity factor for cracking mode α = I, II or III, r and θ are polar coordinates with origin at the crack tip, f_{ij}^{α} is a dimensionless quantity that depends on load, geometry and cracking mode. i and j refer to coordinate axis: i = x, j = y. The term $0\sqrt{\frac{i}{r}}$ becomes negligible when r approaches 0.

The magnitude of K_α depends on sample geometry, the size and location of the crack and the modal distribution of loads on the material. Expressions for f_{ij}^{α} and K_α are available in handbooks and fracture mechanics textbooks.

The critical value of stress intensity factor (K_C) provides a measure of fracture toughness. A fracture (rapid crack growth to failure) occurs when K_α exceeds $K_{\alpha c}$. As the specimen thickness increases, K_α decreases until it reaches a constant value called the plane strain fracture toughness. For a crack in opening mode I, the crack size at the plane strain toughness is called the critical size (a_c), given by the following formula:

$$K_{IC} = \sigma_c Y\sqrt{a_c} \qquad [1.6]$$

where σ_c is the applied stress, Y is a quantity that depends on loading configuration.

Theoretically, K_{IC} is an intrinsic material property that characterizes the resistance to crack or flaw extension. It is an indication of the amount of stress required to propagate a pre-existing flaw, or of the allowable crack sizes under a given load, according to the magic triangle shown in Figure 1.7. The larger K_{IC}, the longer the cracks tolerated by the material. K_{IC} is less than 5 $MPam^{1/2}$ for most brittle materials (Table 1.2). It seldom exceeds 10 $MPam^{1/2}$ for particulate reinforced ceramics or phase transformation toughened zirconia. These values are ridiculously small compared to metals for which K_{IC} exceeds 50 $MPam^{1/2}$ up to 100 $MPam^{1/2}$ for copper and nickel, super alloys, and 170 $MPam^{1/2}$ for steel. For brittle metals such as cast iron and intermetallics, K_{IC} is in the same range as for ceramics. K_{IC} is inappropriate for continuous fiber reinforced composites, because formula [1.5] does not apply, owing to highly heterogeneous microstructures at the crack tip. However, in a first approximation, estimates as large as 40 $MPam^{1/2}$ have been obtained for SiC/SiC composites. In summary, small K_{IC} indicates high sensitivity to cracks and flaws. According to equation [1.6], the critical size of a through-thickness crack ($Y = \sqrt{\pi}$) under a stress of 300 MPa (Figure 1.4) is 56 μm for silicon carbide (K_{IC} = 4 $MPam^{1/2}$) against 35 mm for superalloy (K_{IC} = 100 $MPam^{1/2}$) and 102 mm for steel (K_{IC} = 170 $MPam^{1/2}$).

	K_{IC} (MPa $\sqrt{m}$)
Metals	
Iron	80
Steels	20 – 170
Superalloys	>100
Aluminum	45
Cast iron	5 – 30
Intermetallic alloys	12 – 30
Glass	0.7 – 0.8
Alumina	3 – 5
SiC	4
Si_3N_4	4
Zirconia	4 – 12
SiAlON	5
Concrete	0.2
Limestone	0.9
Ice	0.12
SiC fibres	1.5
Long fiber reinforced SiC *	> 40
Carbon/carbon composites *	notch insensitive
Polymers	
Polyethylene	1 – 5
PTE	3.5
Nylon	3 – 5
Resin	0.6 – 1.0

* K_{IC} is not an appropriate concept for continuous fiber reinforced ceramic or carbon matrix composites. However, it is used for comparison purposes. It refers to an equivalent material. Several continuous fiber reinforced ceramic or carbon matrix composites are insensitive to notches and cracks.

Table 1.2. *Values of K_{IC} for various materials*

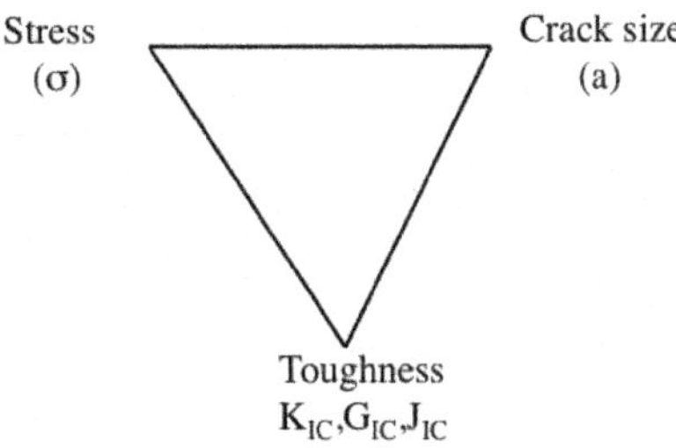

Figure 1.7. *Magic triangle of fracture mechanics relations between stress, crack size and toughness*

Table 1.2 shows a wide family of flaw sensitive materials that includes ceramics, glass, concrete, polymers and some metals. Under uniaxial tensile conditions, cracks are the most severe flaws, owing to a crack tip radius of curvature that is as small as the interatomic distance. However, pre-existing cracks are not the only flaws present in materials, nor the only origin of a fracture. There are many other flaws that differ in shape, dimensions and type.

1.4. The flaws

Flaws are generated during the successive steps of material processing, component manufacturing and service. They can be classified into two families with respect to their size: those flaws with a smaller size than some characteristic dimension of material microstucture (the submicrostructure flaws), and those with larger size (the processing flaws, the machining flaws and the flaws caused by in-service usage).

1.4.1. *Submicrostructure flaws*

These flaws are smaller than the elementary constituents of microstructure, such as grains in the polycrystalline ceramics. They consist of microvoids, inclusions (crystallites), microcracks at grain boundaries or triple junctions (Figure 1.8). Microcracks can be created by thermal stresses during cooling down from the processing temperature. These flaws are the least severe.

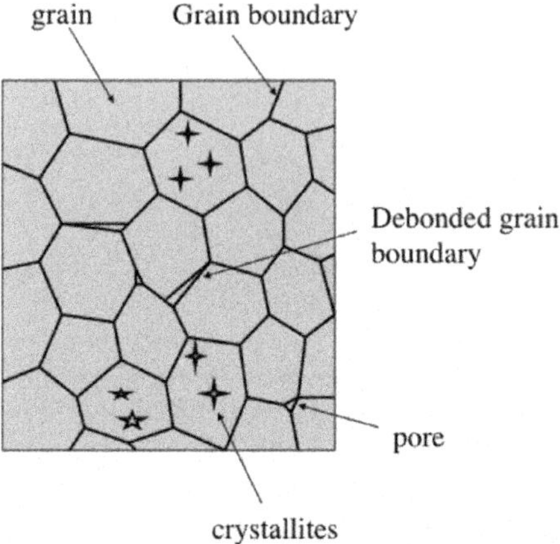

Figure 1.8. *Diagram of the microstructure of a ceramic obtained by powder metallurgy which shows flaws smaller than the grains*

1.4.2. *Processing flaws*

The processing flaws discussed in this section are larger than the characteristic dimension of microstructures (generally the grain size). They can be attributed to process failures or fluctuations. For instance, in sintered ceramics, flaws form during the successive processing steps: milling, mixing, forming, sintering and cooling down from the sintering temperature. Some of the more common processing flaws are agglomerates, pores, voids, inclusions, large grains and cracks (Table 1.3). Figures 1.9–1.11 show the various fracture origin defects detected in broken ceramic test specimens: an agglomerate in a sintered silicon nitride ceramic (Figure 1.9), a metallic inclusion coming from the ball mill during the step of powder milling (Figure 1.10), voids located at the surface or beneath the surface in silicon carbide bending bars (Figure 1.11). It is worth noting some features of these flaws: the size as small as about 100 μm, the shape, the nature and the curvature radius which looks larger than the crack tip one. However, the surface of voids is uneven with some sharp parts at grain boundary junctions. Thus, the curvature radius can be locally smaller than the apparent one. Void severity will depend on local curvature radius and flaw orientation with respect to stress direction.

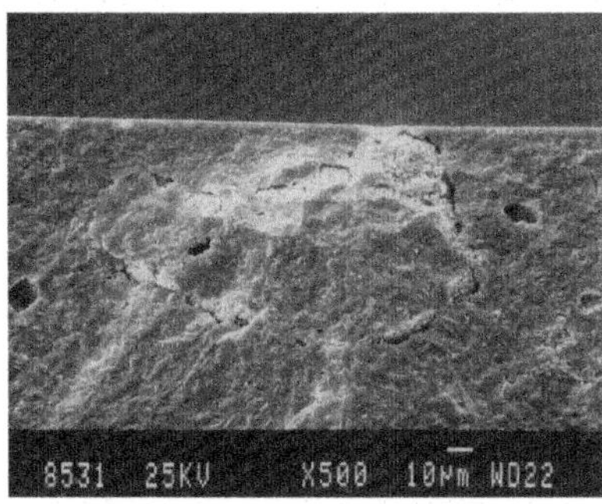

Figure 1.9. *Agglomerate in the fracture surface of a silicon carbide specimen*

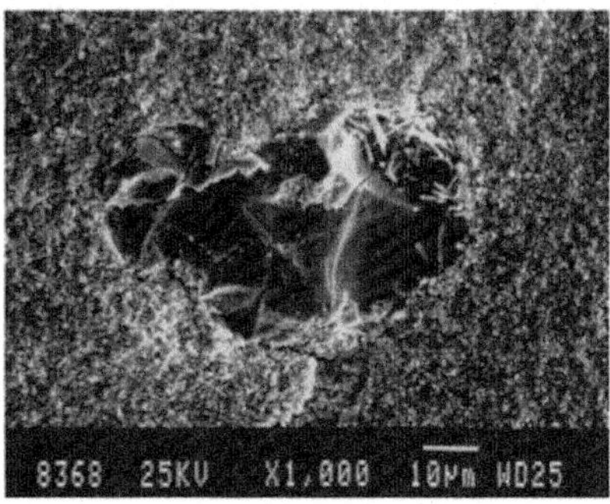

Figure 1.10. *Iron inclusion in the fracture surface of a silicon carbide specimen*

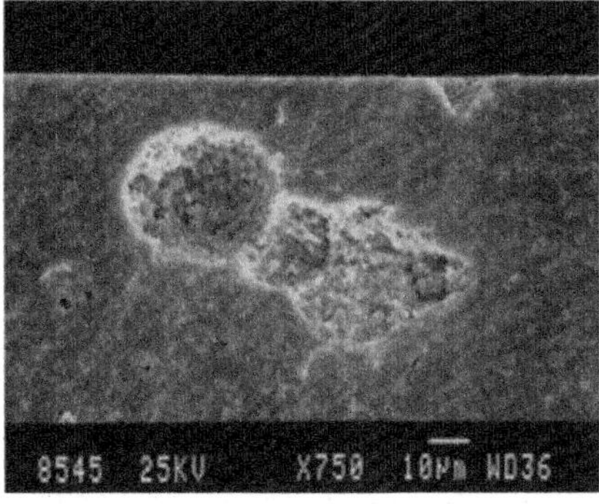

Figure 1.11. *Pore in the fracture surface of a silicon carbide specimen*

Hot pressing	Sintering	Chemical vapor deposition	Fibers from polymer precursors
Inclusions Clusters of large grains Agglomerates	Pores Inclusions Clusters of large grains	Surface flaws Clusters of large grains Fault planes	Pores Voids Chemical heterogeneities

Table 1.3. *Flaws found in ceramics obtained via various processing routes*

In dielectric and electronic ceramics, pores and delamination cracks can form during processing.

In ceramic, glass or carbon fibers, fractures are generally surface-located flaws which are created by scratching or shocks (microcontact flaws) during handling, or by processing flaws such as voids, grains, chemical heterogeneities and contamination surface flaws. Figure 1.12 shows a fracture-inducing pore in a SiC-based fiber (Nicalon grade). Fracture origin is indicated by the visible associated mirror-like zone. It is important to note the submicron size of the pore, which is commensurate with the nanostructure length scale of fiber made up of nanograins of silicon carbide. The pore is thus bigger than the characteristic dimension of nanostructure indicated by grain size (less than 100 nm).

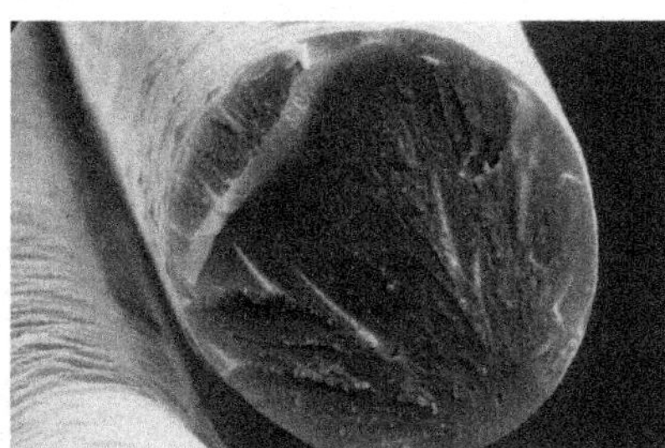

Figure 1.12. *Pore in the facture surface of a silicon carbide fiber (Nicalon NL 202)*

The nature of flaws depends on processing mode, as indicated by Table 1.3 which summarizes the types of flaws identified in ceramics made via hot pressing, sintering, chemical vapor deposition or from polymer precursors.

1.4.3. *Machining flaws*

In ceramics, machining flaws look like scratches, grooves or superficial cracks. They may be created during the finishing step when using diamond tools or abrasive particles. They may form a dual population of flaws of different shapes, with one set of flaws approximately perpendicular and another parallel to the grinding direction. The latter flaws are typically substantially more elongated and often larger than the former. Figure 1.13 shows an example of machining crack that caused failure of a silicon nitride bend bar. The scratch was nearly perpendicular to the stress direction as indicated by the orientation of fracture surface, which corresponds to the most severe configuration.

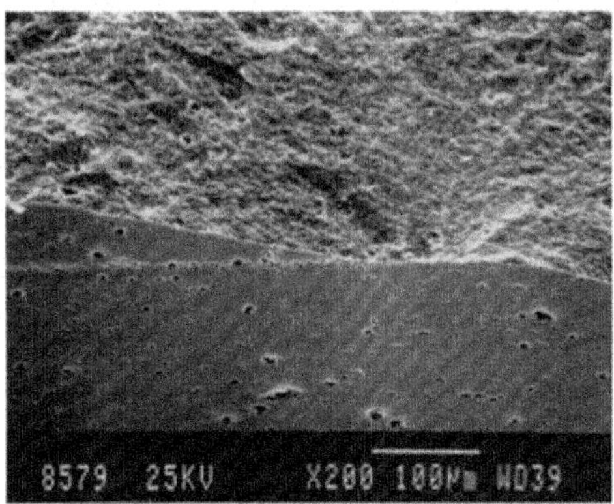

Figure 1.13. *Machining scratch in the surface of a silicon nitride bending specimen which initiated fracture*

Unlike processing flaws which are inherent to material, and as such are intrinsic flaws, machining flaws are extrinsic flaws. They are inherent to the finished piece. They may be present in all the pieces, or only in some of them, depending on the final step of polishing.

More details on machining flaws in ceramics can be found in [QUI 05, RIC 79, RIC 02]

In polymers, a fracture may be induced by micron-sized superficial cracks created during machining or by abrasion.

1.4.4. *Flaws caused by in-service usage*

During service, fracture may occur although the stresses are much lower than the fracture stress. It is initiated by flaws different from the pre-existing flaws, which either formed during service or result from the growth of the pre-existing flaws, under the action of physical or chemical phenomena favored by environmental conditions, such as cavitation during creep at high temperature in oxide ceramics, oxidation in non-oxide ceramics and formation of silicates in silica-based ceramics. Figure 1.14 shows an example of a new fracture-inducing flaw (surface grain-boundary facet) observed in partially stabilized zirconia aged at 1,000°C in air. Figure 1.15 shows a pore blunted by the oxide scale formed in Si_3N_4 during repeated thermal shock heat-up in air. Flexural strength was found to be smaller after aging treatment on partially stabilized zirconia (PSZ), whereas it was slightly larger after thermal fatigue on Si_3N_4. It is worth pointing out that the flaws created during service can be less severe than the pre-existing ones.

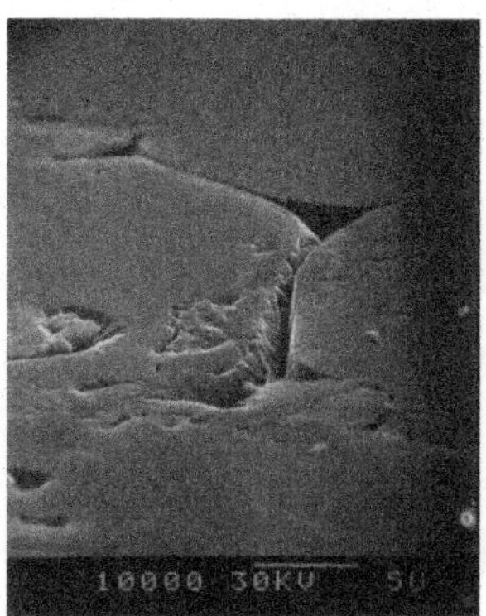

Figure 1.14. *Grain boundary weakening on the surface of a partially stabilized zirconia specimen after aging for 15 h at 1,000°C in air*

1.5. Severity of individual flaws

The severity of flaws depends on flaw characteristics (dimensions, nature and shape), orientation with respect to stress direction and stress field. It is commensurate with stress amplifying capacity. The most severe ones are those for which the intrinsic strength is reached under the smallest remote stress.

1.5.1. *Severity of cracks and voids*

For cracks and voids, flaw severity can be measured using the stress concentration factor K_t (equation [1.3]), which takes into account the shape through the size and the curvature radius at flaw tip. The stress intensity factor K_I is restricted to cracks, and it involves the crack size only. It is a common practice to consider flaws as cracks and to use the linear elastic fracture mechanics approach to design critical components. This approach uses the flaw size and fracture toughness to evaluate the ability of a component containing a flaw to resist fracture (Figure 1.7). A related concept is the strain energy release rate, $G = dU/dA$, where U is the strain energy and A is the crack area. The strain energy release rate increases with load and crack size. If the strain energy release rate exceeds critical value G_c, the crack will grow spontaneously. However, G is generally not formulated in terms of initial crack size, and G_c data are not documented satisfactorily, although G_c can be equated to the surface energy of the two new crack surfaces.

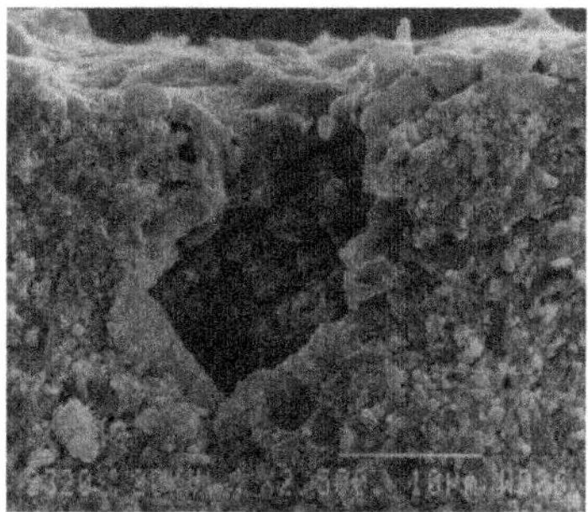

Figure 1.15. *Pore coated by a layer of oxide in a silicon nitride specimen after repeated thermal shock heat-ups (1,000 cycles) in air*

In a uniform uniaxial tensile stress field, the severity is commensurate with flaw size, at first approximation. However, depending on flaw shape, the orientation to stress direction may affect this criterion. For cracks of the same size, G and K_I decrease when the angle of the crack plane to stress direction decreases to zero. The stress concentration factor K_t relates stresses perpendicular to the principal flaw plane, then equation [1.3] requires appropriate stress components.

For different flaws with the same characteristic size (in the plane perpendicular to the direction of propagation), which propagate perpendicular to the stress direction (uniform uniaxial tensile stress field), cracks are more severe than voids and inclusions. According to the stress concentration factor K_t, flaw severity is commensurate with curvature at flaw tip ($1/\rho$).

Flaw severity also depends on the stress state. In a non-uniform stress state, the intrinsic strength can be reached near a flaw although it is small or has a large curvature at tip, when it is located in the region where high stresses operate. This flaw would not be critical in a uniform stress-state. As before, severity is enhanced by flaw orientation with respect to stress direction. The most severe flaws are perpendicular to the stress direction. Flaws that would not be critical under uniaxial tension can be severe in a multiaxial stress field when they are perpendicular to one stress direction.

In summary, it is obvious that the most severe flaw is the one which will cause fracture. Severity results from the combination of several factors including the characteristic size, the nature, the orientation and the location. Flaw severity is appropriately characterized by the stress concentration factor, or the stress intensity factor in the above particular cases. In the general case, strain energy release rate can be an appropriate alternative. The local stress is also an interesting concept, as discussed in the chapter on probabilistic approaches based on the elemental stress concept.

Severity of inclusions is a particular case since the above concepts K_t, K and G do not apply. On the other hand, the local stress concept can be used.

1.5.2. *Severity of inclusions*

The severity of inclusions depends additionally on the residual stresses induced by expansion mismatch during processing or service. When the residual stresses exceed the intrinsic strength, they induce local damage or fracture. For the materials made at high temperature, the residual stresses build up during cooling down from the processing temperature. They are determined by the respective thermoelastic properties of inclusion and surrounding material, i.e. the coefficient of thermal expansion and elastic modulus. Damage and fractures are governed by the strengths of inclusion, interface and material. The following different cases can be observed during cooling down from the processing temperature on brittle materials like ceramics:

– when an inclusion has a smaller coefficient of thermal expansion compared to material, it hampers the contraction of surrounding material, which generates tensile stresses in the material and compressive stresses in the inclusion. If the tensile stresses exceed material intrinsic strength, a crack forms the size of which depends on stress gradient. This crack will be a pre-existing flaw of the material. Its severity will depend on its orientation to the direction of applied stress;

– instead, when an inclusion has a large coefficient of thermal expansion compared to material, it contracts more than the surrounding material, which generates tensile stresses in the inclusion and in the interface, and compressive stresses in the material. If the inclusion/material interface is weak compared to the tensile stress that is generated, debonding will occur, which generates a void, which will act as a pre-existing flaw. If the inclusion/material bond is strong, the inclusion is the weakest link. It will crack if its strength and toughness are small. The crack in the inclusion will be a pre-existing flaw that may be innocuous depending on interface crack arrest capability;

– finally, if no crack forms during the processing step, the presence of the inclusion affects deformations in the surrounding material during further loading, and generates stress concentrations. The location of stress concentrations is dictated by the respective Young's moduli of the inclusion and the material. When the inclusion is stiffer, stress concentrations are localized near the poles of inclusion (the poles are located in the plane parallel to loading direction; Figure 1.16). Alternately, when the Young's modulus of inclusion is small, stress concentrations occur in the equatorial plane (the plane perpendicular to the loading direction; Figure 1.16).

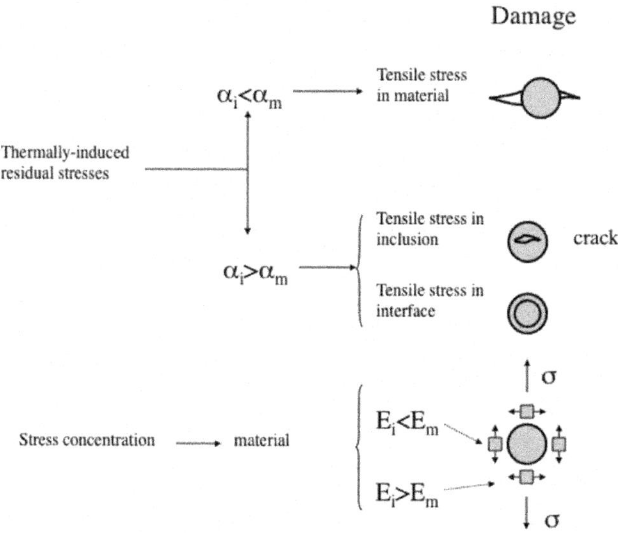

Figure 1.16. *Effects of inclusions on surrounding material: thermally induced residual stresses in the surrounding material that may cause cracks or voids, or stress concentration under load in the surrounding material depending on respective elastic moduli of inclusion and surrounding material*

Failure studies reveal an appreciable sensitivity of the failure condition to the defect type and to the surrounding microstructure in polycrystal ceramics [EVA 80, MEY 80, FER 80]. Experimental data shown in Figure 1.17 reveal the more or less significant influence of various types of inclusions on the fracture strength of a silicon nitride ceramic. Silicon inclusions seem to be more severe than the carbon and iron inclusions, whereas the tungsten carbide ones look innocuous. Figure 1.17 also indicates the ranking of the severity of different types of flaws in a polycrystal ceramic.

Determination of the stress field in and around an inclusion and an inhomogeneity has been a focus of solid mechanics for several decades, since the celebrated work of Eshelby [ESH 57, ESH 59, ESH 61]. Detailed studies have been conducted for a perfectly bonded or imperfectly bonded inhomogeneity embedded in an infinite or semi-infinite domain. Stress formulas for the region outside an arbitrarily oriented spheroidal inhomogeneity in an infinite matrix under a remote axisymetric loading have been established. The effect of inclusions with particular shapes can be investigated using numerical methods.

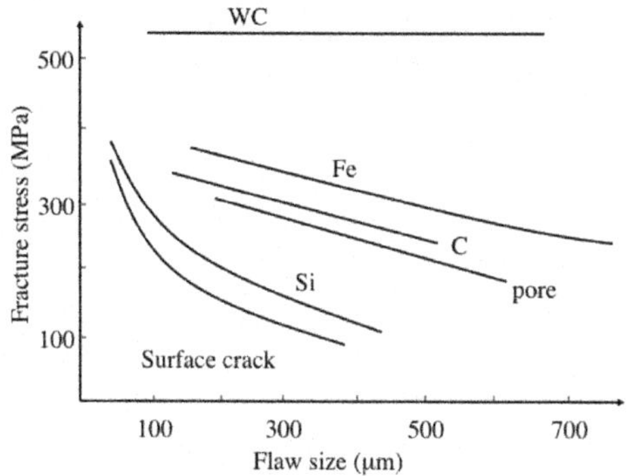

Figure 1.17. *Influence of various types of flaw on the fracture stresses of silicon nitride specimens (from [EVA 80])*

1.6. Influence of flaw populations

In many materials like polycrystal ceramics, flaws are not unique, but instead they form one or several populations. As indicated before, the severity of each flaw depends on at least five parameters. Thus, it is impossible that two samples of the same material, identical in size and shape, possess identical flaws at any location. In the case of a single population, the flaws should differ by their characterististic size, their orientation. In the presence of multimodal populations, they also differ by their nature, as shown in Figure 1.17. Thus, characteristics of the critical flaw vary from specimen to specimen. In the simplest case of a unique flaw population and applied uniaxial tensile stress, flaw criticality depends on characteristic size and orientation with respect to stress direction. In a non-uniform stress state, flaw location is an additional parameter. In the presence of several populations, flaw nature is again an additional parameter. Thus, flaw criticality is a random variable.

It is implicitly considered that the distribution of flaws is homogeneous, which is a reasonable assumption for well-developed engineering materials.

1.6.1. *Strength variability*

The fracture strengths measured on a batch of identical specimens of the same material display a significant scatter which reflects the features of flaw criticality. Figure 1.18 plots the fracture strengths obtained on several identical bending bars of the same ceramic. Under the same loading conditions (either 3-point or 4-point bending), the strength interval is as large as half the minimum strength. Much wider intervals can be obtained, for instance on ceramic fibers, for which they may be as large as five times the minimum strength. It is worth pointing out that these intervals are much larger than those that would result from experimental uncertainty. The strength interval reflects variability in critical flaw severity. The fracture strength is a statistical variable. It takes discrete values that differ for random causes and, when ranked in ascending order, follow a statistical distribution or array. The cumulative distributions of strengths shown in Figure 1.18 describe the probability that the real-valued random variable strength will be found at a value less than or equal to S. The method used to obtain the Weibull plot shown in Figure 1.18 is described in Chapter 7. As discussed in Chapter 7, this logarithmic plot corresponds to the linearized equation of cumulative distribution. It is commonly used. It is convenient to show the presence of several populations of fracture-inducing flaws. It is linear in the presence of a single population.

Figure 1.18 shows that a certain amount of specimens exhibit strengths weaker than a given stress: for instance, 18 of 70 specimens for 1,000 MPa in 3-point bend (Figure 1.18(b)). This means that 18 of the 70 specimens were unable to carry the stress of 1,000 MPa. Thus, when one of these specimens was drawn for testing, it had 18 in 70 chances of fracturing under 1,000 MPa. It was impossible to predict the behavior of the selected specimen using the fracture toughness–crack length formula (equation [1.6]). A fracture is an event whose outcome cannot be predicted. Under given conditions, it may or may not occur. Fracture is a random event. Under a given stress, it is governed by the probability of presence of a critical flaw.

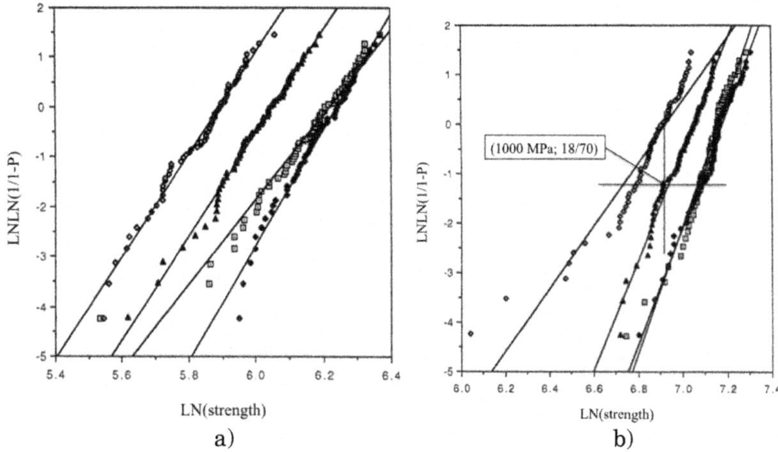

Figure 1.18. *Linearized cumulative distributions of fracture strengths measured under various flexure conditions: (◆) short-span (SS) 3 pt-bending tests, ■ intermediate-span (IS) 3 pt-bending tests, ▲ long-span 3 pt-bending tests, (◇) 4-point bending, for silicon carbide a) silicon nitride b)*

1.6.2. *Size effects*

The fracture strength of brittle materials depends on specimen size. The larger the size, the smaller the failure strength. This effect is related to the probability of presence of a critical flaw. The larger the specimen, the higher the probability of presence of large flaws. This effect is shown by the comparison of ceramic strengths measured on centimeter size specimens (about < 800 MPa) and micrometer diameter fibers (about 1–3.5 GPa, Table 1.1). The critical flaw size in SiC carbide bend bars is around 100 µm, whereas it is about 300 nm in SiC-based fibers having 12 µm diameter (section 1.3.2). The size effect on strength is also shown in Figure 1.18, which shows that the failure strengths of the small specimens (i.e. the intermediate span (IS) and short span (SS) specimens that were, respectively, 10 and 8 mm long) were higher than that of the large specimens (the other specimens with 40 mm length).

1.6.3. *Influence of stress field*

As the probability of presence of a critical flaw is commensurate with the size of stressed volume, the trend in the influence of stress field is similar to size effect discussed above, when the volume under high stress increases. This is observed in uniaxial stress states with gradients. Figure 1.18 shows that the bars tested in 3-point bends broke under high stresses when compared to identical bars loaded in 4-point bending conditions. The 3-point bending bars experience a stress gradient, while uniform stresses operate in a significant part of the 4-point bending specimens. As a result, the volume under high stresses is larger in the 4-point bending specimens.

The relation between the stress state and the stressed volume involves the peak stress. It is shown in Chapter 6 that this relation is valid in most cases, even if the stressed volume may stay constant when the peak stress increases.

Enhanced influence of stress state is highlighted when comparing the strengths of tensile and 3-point bending specimens (Figure 1.19). The entire volume of tensile specimens is subjected to the peak stress. By contrast, the 3-point bending specimens failed from the outer surface. The amount of stressed material was larger in the tensile specimens. From Figure 1.20, it can be noted that similar effect is observed in polymer matrix composites reinforced by glass fibers, for which the strengths rank in the same increasing order when comparing tension, 4-point and 3-point bending specimens.

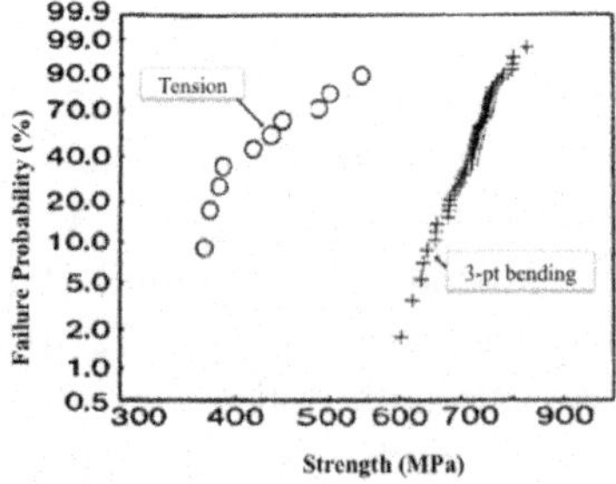

Figure 1.19. *Linearized cumulative distributions of fracture strengths measured in tension and in 3-point bending for silicon nitride (from [KAW 86]). (Data are plotted as LnLn ($1/_{1-P}$) vs. Ln σ_{max} (Chapter 7))*

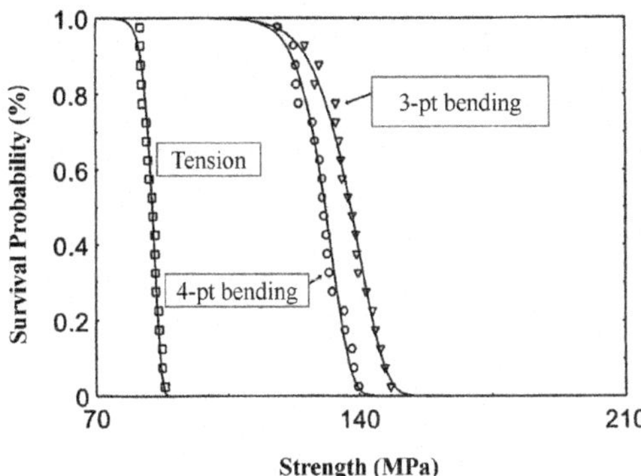

Figure 1.20. *Survival probability versus fracture strengths measured in tension and in 3-point and 4-point bending for a laminated polymer matrix composite (from [PRA 99])*

In multiaxial stress-states, a larger amount of flaws are activated, as a result of random orientation of flaws. As a consequence, strength decrease is expected when comparing with uniaxial stress states, even under constant stressed volume. This issue is discussed quantitatively in Chapters 5 and 6.

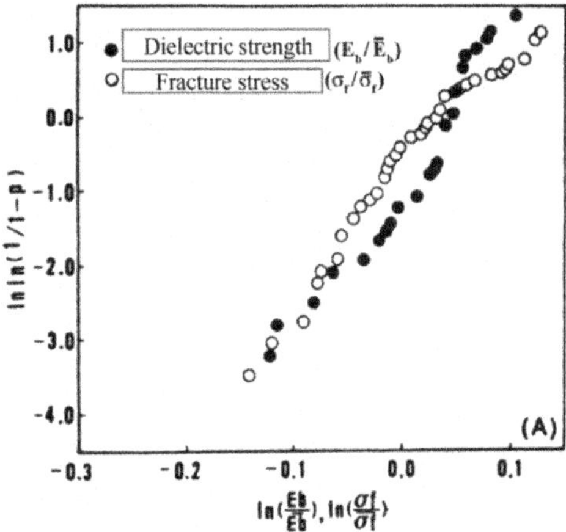

Figure 1.21. *Linearized cumulative distributions of dielectric strengths for $BaTiO_3$ (from [YAM 84])*

1.6.4. *Influence of loading conditions*

The stresses can also be generated by other loading modes, such as thermal shock and electric field. Fracture is still a random event as shown by the cumulative distributions of strengths (Figures 1.21 and 1.22) characterized by values at failure of applied electric field or of temperature difference between the material and environment. In both cases, failure is caused by a critical flaw. The strength is sensitive to stress-state and flaw population.

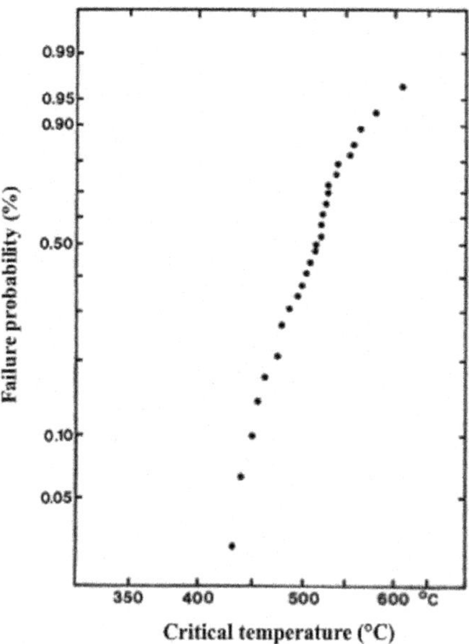

Figure 1.22. *Linearized cumulative distributions of thermal shock resistance (in terms of critical temperature difference) for alumina specimens subjected to quenching thermal shocks. ((Data are plotted as LnLn ($1/_{1-P}$) vs. Ln ΔL_C (Chapter 7))*

A thermal shock generates transient stresses. Time-dependent temperature and stress gradients develop in the material. During cooling down, the temperature decrease starts from specimen surface, then, the cold front moves to the interior. Tensile stresses are generated first in the surface that is prevented from contracting by the interior of specimen still at initial temperature. Then, tensile stresses build up in the rest of specimen as temperature decreases. Fracture does not occur at the very beginning despite the presence of high tensile stresses in the surface, but, instead later when tensile stresses operate on a sufficient volume of specimen (Figure 1.23). Failure occurs when one flaw becomes critical which requires that a sufficient volume of material is subjected to a certain level of stresses. The resistance to failure by thermal shock is also sensitive to size effects (Figure 1.24). When the size of the zone or the specimen exposed to cooling down is small, the chances of presence of a critical flaw decrease, and the resistance to failure is high (Figure 1.24).

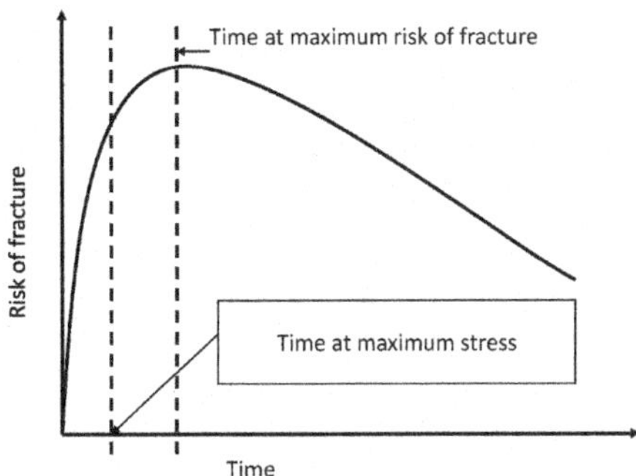

Figure 1.23. *Risk of fracture during a quenching thermal shock*

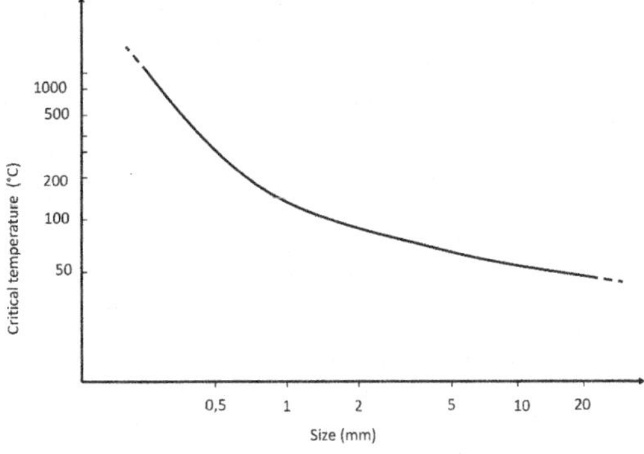

Figure 1.24. *Size effect during a quenching thermal shock: dependence of the thermal shock resistance on the diameter of the cooled down part of specimens*

1.6.5. *Influence of multimodal flaw populations*

Materials may possess bimodal or multimodal flaw populations, which are different by the location (surface-located and volume-located flaws), nature (pores, cracks and inclusions), origin (extrinsic and intrinsic flaws) or other features (weak fibers and strong fibers in fiber bundles) of the flaws. However, for fracture strength issues, severity is the key feature that differentiates the flaws. There are examples of surface-located and volume-located flaws which display the same severity.

The presence of multimodal flaw populations can enhance size effects on fracture strength, when the volume change favors the contribution of one flaw population. Figure 1.25 shows the case of a bimodal population of surface-located and volume-located flaws, when the smaller strengths result from volume-located flaws. This situation is generally observed on massive specimens for which the smaller strengths result from. The cumulative distributions of fracture strengths show that the contribution of volume-located flaws increases with specimen size. The transition strength S* was taken to be constant. The trend would be more significant with size-dependent S* (see Chapter 6).

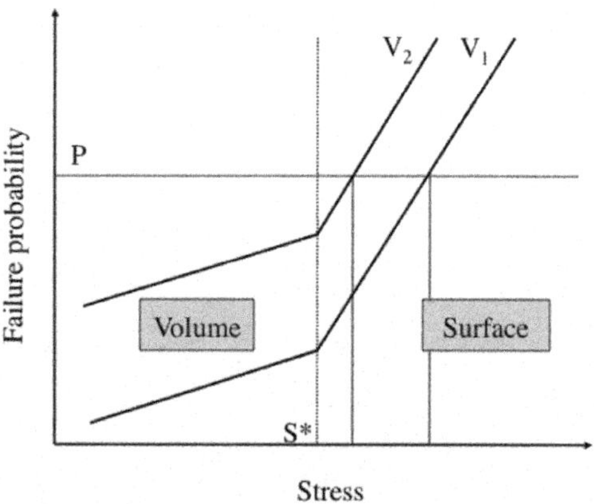

Figure 1.25. *Size effect on strength: influence of volume of specimens ($V_1 < V_2$) on linearized distributions of fracture strengths in the presence of bimodal flaw populations (surface and volume located flaws)*

The stress-state may also favor the contribution of one flaw population. Figure 1.19 shows this effect using experimental data obtained in tension and in 3-point bending test on ceramic. The linearized cumulative distributions of strengths have a different slope, which reflects the presence of different flaw populations (this question is discussed in Chapter 7). Since all the flaws are loaded uniformly in the tensile specimens, and high stresses operate in the outer surface of 3-point bending specimens, it can be concluded that volume-located flaws in tension, and surface-located flaws in 3-point bending dominated fracture.

1.6.6. *Effects of environment*

Environment may cause change or growth of pre-existing flaws, or formation of new flaw populations. Thus, under stresses smaller than material strength, fractures occur when one of the flaws reaches a critical size. Flaw severity time-dependence is dictated by the kinetics of environment-controlled reactions. The effect of environment may be activated by stresses.

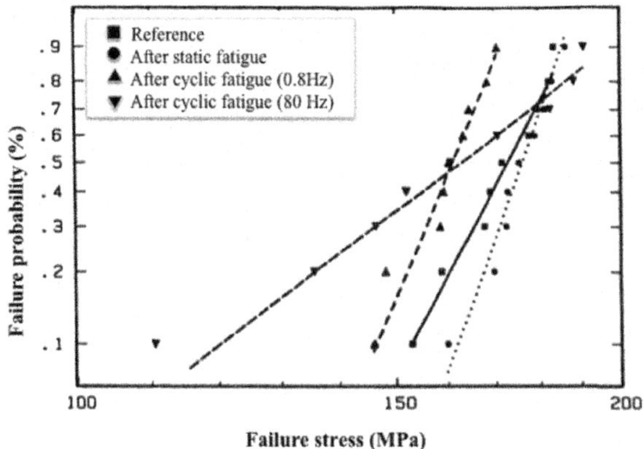

Figure 1.26. *Influence of static and cyclic fatigue at 1,200°C in air (during 7 h) on the linearized distribution of room temperature tensile strengths for silicon nitride*

Figure 1.26 shows the effect of fatigue at high temperature in air on the cumulative distributions of room temperature strengths of silicon nitride specimens. Strength degradation was measured after the cyclic fatigue tests, and the slope of the linearized cumulative distribution after high-frequency fatigue decreased as well. These features reflect the presence of more severe flaws after cyclic fatigue at high temperature, and different populations depending on frequency. It was found that high-frequency loading enhanced oxygen penetration into the material, which affected the population of internal flaws. On the contrary, low frequency affected the surface defects, which resulted in a larger slope for the distribution of strengths.

By contrast, it can be noted that fracture strengths were improved after the static fatigue tests at high temperature. Flaws were less severe, as a result of blunting due to the oxide scale observed at the surface of specimens. Flaw blunting was ineffective during cyclic fatigue because the oxide layer was degraded. Similar strength increases have been observed on ceramics after corrosion in very aggressive media at high temperature. Environment can thus enhance or reduce the severity of flaws.

1.7. Consequences of failure predictions

There are several approaches to deal with the influence of flaws on structural integrity and engineering application of materials:

– according to common practice, consider flaws as cracks and use the linear elastic fracture mechanics approach to design critical components. However, this approach neglects the numerous effects induced by the flaws;

– produce flaw-free materials. But, it is an utopia since it is a fact of life that any material contains flaws;

– accept materials with flaws, and employ appropriate methods to design components.

As shown in this chapter, the fracture of ceramics displays numerous features that may be summarized as follows and that must be taken into account. A fracture is induced by flaws that are either intrinsic or extrinsic, and which form populations generally homogeneous over the material. The flaws have a random distribution.

Flaw severity depends on several factors including characteristic size, shape, nature, orientation and location and stress state. Brittle fracture is a random event and fracture strength is a statistical variable. Fracture strength depends on geometry, dimensions and loading conditions. It is sensitive to size and stress-state effects that may be enhanced in the presence of multimodal flaw populations. As a result, strength data have a relative value only. They cannot be used as such for fracture predictions.

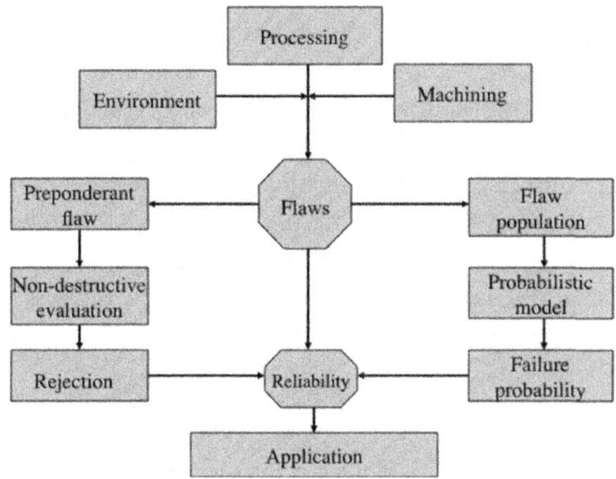

Figure 1.27. *Approaches to ceramic reliability based on either the non-destructive detection of the biggest flaws or the characteristics of the flaw population*

Non-destructive methods can be used to identify the flaws (Figure 1.27). These methods are suitable to detect big individual flaws, but they are unable to evaluate flaw severity. Destructive methods are used to determine material strengths. Statistical-probabilistic approaches provide the relations between the three fundamental factors governing fracture: characteristics of flaw populations, stresses and fracture probability. According to this triangle, any factor can be derived when the two others are known (Figure 1.28). Of particular interest is the stress as a function of probability of fracture, or probability of fracture as a function of applied stress. This triangle includes the deterministic fracture mechanics triangle at flaw/crack scale that relates toughness, stress and crack size (Figure 1.7). The latter has a local merit as it can be applied to a single crack or flaw, whereas the former has a global value as it applies to the whole material. The probabilistic triangle represents the probabilistic approaches that consider the flaws as physical entities, which are discussed in this book. These approaches are developed as follows: the criterion of local fracture from a flaw is established first, then the probability of presence of this flaw is expressed as a function of the stress-state and characteristics of flaw strength distribution. We may propose an analogy with statistical thermodynamics that describe the behavior of systems containing a large number of particles. Statistical thermodynamics is based on the fundamental assumption that all possible configurations of a given system, which satisfy the given boundary conditions such as temperature, volume and number of particles, are equally likely to occur. The overall system will therefore be in the statistically most probable configuration. From a fracture point of view, a material is a system containing a large number of flaws which have any location, and a severity dependent on location.

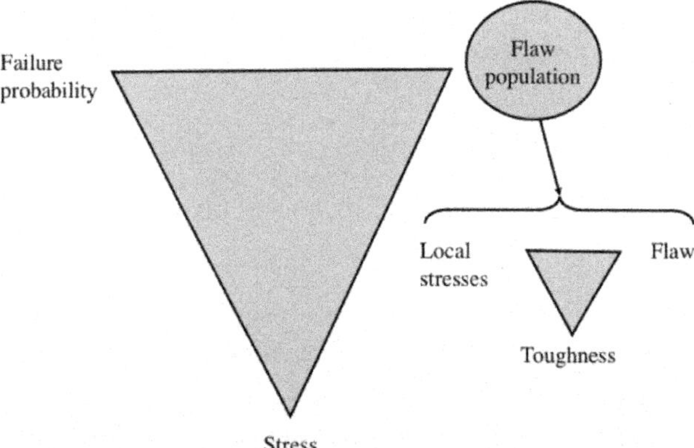

Figure 1.28. *Magic triangle of probabilistic fracture mechanics showing relationships between stress, fracture probability and flaw population. The magic triangle of fracture mechanics applies locally at flaw level*

The statistical-probabilistic approaches can be used to predict the fracture stress of specimens or components containing flaws. They allow the interpretation of strength variability. Moreover, they should enable the use of laboratory data to infer the probability of fracture of a component under more complicated stress-states encountered in service (non-uniform polyaxial stress-states and transient stress-states), which requires considerable extrapolation to larger sizes as well as to more complex stress-states than those encountered with specimens usually tested under uniaxial stresses. An appropriate approach to brittle failure is an essential prerequisite to a sound failure prediction.

There are two main types of probabilistic-statistical approaches to brittle fracture:

– the phenomenological and macroscopic approaches. The Weibull model is famous;

– the aforementioned approaches that consider the flaws as physical entities. They are more fundamental. They are based on flaw strength density function. Severity is measured either using an elemental strength or flaw size.

The statistical-probabilistic approaches to fracture are based on the weakest link concept which assumes that fracture of the bulk specimen occurs instantaneously as soon as the crack initiated. Brittle material fails like a chain, the links of which would be specimen volume elements. The chain model can be extended to damage by progressive multiple cracking observed in some materials such as multilayers and continuous fiber reinforced composites. This issue is discussed in Chapter 10. It must be stressed that care must be taken not to violate the weakest link assumption, but to use the probability equations properly. The "weakest link" concept assumes that fracture of the bulk specimen is determined by the strength of its weakest volume element. This implies that the weakest link concept is restricted to unstable fracture by direct extension of crack throughout the bulk of the specimen. The mechanisms of stable propagation or progressive coalescence of cracks are discounted.

2

Statistical-Probabilistic Approaches to Brittle Fracture: The Weibull Model

2.1. Introduction

The name Wallodi Weibull (1887–1979) is associated with the domain of statistics on life data of systems and analysis of failures. Weibull was interested in problems of material strength, such as fatigue and failure of vacuum tubes. In 1939, he proposed a theory based on the elementary laws of probability, whose formulae may solve inconsistencies observed within the classical theory of the ultimate strength of a material. He introduced the Weibull distribution function that was then discussed in his paper of 1951 which claimed that this statistical distribution is of wide applicability. Weibull illustrated this point with various examples including the yield strength of steel, fiber strength of Indian cotton, fatigue life of steel, statures for adult males, etc. This distribution was first identified by Fréchet (1927) and first applied by Rosin and Rammler (1933).

The Weibull distribution is the most widely used for fitting and analyzing life data. According to Weibull, "the only merit of this distribution function is to be found in the fact that it is the simplest mathematical expression of the appropriate form, which satisfies the necessary general conditions. Experience has shown that, in many cases, it fits the observations better than other known distribution functions".

However, the Weibull distribution does not always work [LAM 88b]. It presents shortcomings for fracture problems. A major criticism is that the Weibull theory makes no attempt to incorporate a physical description of the fracture-initiating flaws, and does not represent a fundamental way of treating the multiaxial effect. The Weibull approach is a phenomenological and macroscopic approach, as opposed to the more fundamental statistical approaches to brittle failure that recognize the flaws as being unique entities. These features imply limits to fracture predictions.

E.J. Gumbel showed that the Weibull distribution and the type III smallest extreme value distributions are the same. He also proved that if a part has multiple failure modes, the time to first failure is best modeled by the Weibull distribution. Extreme value theory is the theory of modeling and measuring events which occur with very small probability. In the general literature of mathematics and statistics, the Weibull distribution is cited as the type III standard distribution function. Type I is the Gumbel distribution function, and type II is that of Fréchet.

In this chapter, the Weibull distribution is established from a fracture point of view. In Chapter 5, it is shown that the Weibull equation is a particular case in fracture problems. Various conditions of stress-state (uniform, non-uniform and multiaxial) and fracture location (surface and volume) are examined.

2.2. Weibull statistical model

2.2.1. *The weakest link concept*

First, the material is modeled as a chain of n links. The chain model was based on the observation that the breaking load is not exactly the same when identical rods are ruptured by an external force. The failure stress of the chain is determined by the strength of its weakest element. The relationship with a material was made as follows. Weibull considered that a body contains a statistical distribution of non-interacting (physically unspecified) inhomogeneities. The material is subdivided in volume elements which contain each a single inhomogeneity. The volume elements compete for failure as the links of a chain do. The winner is the volume element that causes fracture. This is the one with the inhomogeneity of critical severity. In short, a link in the chain corresponds to a volume element.

Second, it is assumed that the strengths of links are independent variates.

Third, the distribution of link strengths is homogeneous over material volume.

2.2.2. *Probability of fracture under a uniaxial tensile stress*

The condition for fracture is based on the definition of the probability of independent events occuring: when two events are independent, the probability of both occurring is the product of probabilities of each occurring. The events of simultaneous non-failure of the links in a chain are independent. Unlike fracture, the fact that a link survives does not affect the probability of survival of another link.

The survival probability of a volume element V under stress σ is:

$$P_s\,(V, \sigma) = P_v\,(\sigma_R > \sigma) \qquad [2.1]$$

σ_R is the element strength.

For two elements with volumes V and V', the survival probability of volume V+V' under the stress σ:

$$P_s\,(V+V', \sigma) = P_s\,(V, \sigma).\ P_s\,(V', \sigma) \qquad [2.2]$$

This condition is satisfied by an exponential function, since the function must transform a product into a sum and vice versa. This is demonstrated as follows. Taking the logarithm of expression [2.2], it follows that:

$$\text{Ln}\ P_s\,(V+V', \sigma) = \text{Ln}\ P_s\,(V, \sigma) + \text{Ln}\ P_s\,(V', \sigma) \qquad [2.3]$$

Then, expression [2.3] is differentiated with respect to variable V at constant V':

$$\frac{d}{dV}\,\mathrm{Ln}\ P_s\,(V+V', \sigma) = \frac{d}{dV}\,\mathrm{Ln}\ P_s\,(V, \sigma) \qquad [2.4]$$

The expression [2.4] is true ∀V or V'. As a result, it depends on σ only.

$$\frac{d}{dV}\,\mathrm{Ln}\ P_s\,(V, \sigma) = \phi(\sigma) \qquad [2.5]$$

Integrating equation [2.5], it follows that:

$$P_s\,(V, \sigma) = \exp\,[-V.\ \phi(\sigma)] \qquad [2.6]$$

The probability of fracture of volume V is:

$$P\,(V, \sigma) = 1 - P_s\,(V, \sigma) = 1 - \exp[-V.\phi(\sigma)] \qquad [2.7]$$

The necessary general condition that the function $\phi(\sigma)$ has to satisfy is to be a positive, non-decreasing function so that the probability of fracture increases with the applied stress. According to Wei-bull, the most simple function satisfying this condition is:

$$\phi(\sigma) = \left(\frac{\sigma - \sigma_u}{\sigma_o}\right)^m \qquad [2.8]$$

As m infinitely increases, the strength value derived from the expression of probability of fracture tends toward σ_0, which is consistent with the classical theory of fracture according to which fracture occurs when the internal stresses in any point of the body, irrespective of its dimensions, have reached a critical value. Then, the standard deviation tends toward 0 when m infinitely increases.

The justification given by Weibull in adopting this power law representation was that experience has shown that in many cases, it fits the observations better than other known distribution functions.

The probability of fracture corresponding to a uniform stress-state is generally written as:

$$P(V, \sigma) = 0 \text{ for } \sigma \leq \sigma_u$$

$$P(V, \sigma) = 1 - \exp[-V.(\frac{\sigma - \sigma_u}{\sigma_o})^m] \text{ for } \sigma > \sigma_u \qquad [2.9]$$

It is important to point out the following features of the stress function $\phi(\sigma)$. In the 1939 paper, it was denoted by $n(\sigma)$ and referred to as the material function, i.e. a function expressing the strength properties of the material. It represents the distribution function of tensile strengths of test specimens with given size and a certain amount of heterogeneities. Indeed, the bulk strength depends on the material microstructure which includes the flaws. However, the relationship with the microstructure was unspecified and it remains vague. The orientation of flaws with respect to the load direction is an important factor as it contributes to critical severity. In the typical case of uniaxial uniform tension, it can be anticipated that the specimen strength corresponds to those flaws with a preferred critical orientation with respect to the load direction. A different result would be obtained if the test specimens were subjected to different loading conditions. For instance, biaxial loading would activate heterogeneities with different orientations, which would not be critical under uniaxial tension. These remarks imply that the nature of inhomogeneities must be physically considered. This issue is addressed in subsequent chapters.

The power law is a satisfactory approximation of the distribution of the smallest tensile strengths. It can be used to represent the cumulative distribution of the most severe flaws in materials as shown in subsequent chapters. Furthermore, it is a simple expression, which can be handled relatively easily.

2.2.3. *Statistical parameters*

m is a dimensionless parameter called the Weibull modulus. It is an index of the degree of scatter in measured strength values. The scatter decreases with increasing m (Figure 2.1). The following m values have been estimated:

– engineering ceramics: 2–20;

– ceramic matrix composites(*): ≈ 30;

– ceramic fibers: 5–10;

– glass fibers: ≈ 5;

– steels: 50–100.

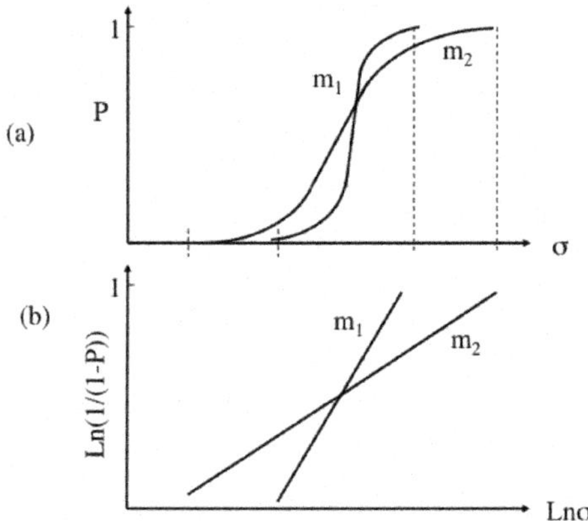

Figure 2.1. *Statistical distributions of strength data for various values of Weibull modulus ($m_2 < m_1$) showing that Weibull modulus varies in inverse proportion to width of strength interval; a) P-σ plot (equation 2.11), b) LnLn(1/1-P) versus Lnσ plot (see Chapter 7)*

(*) For continuous fiber reinforced ceramic matrix composites, the Weibull theory does not apply to the description of ultimate fracture since the weakest link concept is violated. m is just an index of scatter when the cumulative distribution of ultimate strengths is considered.

Weibull assumed that m, σ_0 and σ_u are the material parameters depending only on the microstructure. Batdorf and Lewis III's works, as well as recent results on the fracture of ceramic fibers, indicate that m is a function of specimen size.

σ_0 is a normalized factor, also referred to as scale parameter, or distribution parameter. It is commensurate to the mean strength according to the following equations.

when $\sigma_u = 0$ $$\sigma_0 = \bar{\sigma}_R \frac{V^{\frac{1}{m}}}{\Gamma(1+\frac{1}{m})}$$ [2.10]

when $\sigma_u > 0$ $$\sigma_0 = (\bar{\sigma}_R - \sigma_u)\frac{V^{\frac{1}{m}}}{\Gamma(1+\frac{1}{m})}$$

where Γ (.) is the gamma function: $\Gamma\ (1+\frac{1}{m}) = \frac{1}{m}!$

The scale factor has dimension of the product $\sigma V^{1/m}$ (m is the Weibull modulus). When introducing a reference volume V_0 into equation [2.9], the scale factor has the dimension of σ

$$P(V,\sigma) = 1 - \exp\left[-\frac{V}{V_0}\left(\frac{\sigma - \sigma_u}{\sigma_0}\right)^m\right]$$ [2.11]

V_0 and V must have the same volume unit. When m^3 is used in accordance with the international system, V_0 is set to 1 m^3. V_0 is set to 1 mm^3 with V expressed in mm^3. The value of V_0 must be specified for a proper further application of statistical parameters. If V_0 is not specified, the estimate of σ_0 is meaningless; it cannot be used for safe failure predictions. It is important to stress that the value of σ_0 depends on volume unit. Substantially different σ_0 estimates are obtained if V_0 is set to 1 m^3 or 1 mm^3 as shown by equation [2.12] derived from equation [2.11]:

$$\sigma_0(m^3) = \sigma_0(mm^3)\ \{V_0(mm^3)/V_0(m^3)\}^{1/m} = \sigma_0\ (mm^3)\ 10^{-9/m}$$ [2.12]

where $\sigma_0(m^3)$ and $\sigma_0(mm^3)$ are the estimates corresponding, respectively, to V_0 =1 m^3 and V_0 =1 mm^3.

σ_u is the low strength limit in the three-parameter distribution (equation [2.9]), i.e. the stress at zero probability of failure. However, it is considerably difficult to obtain σ_u from experimental data. As a result, a better description of fracture cannot be expected by using the three-parameter distribution. Furthermore, it is anticipated that the estimates of σ_u would be obtained with an uncontrolled uncertainty. Thus, in most analyses, σ_u is taken to be 0. In the following, $\sigma_u = 0$.

2.3. Probability of fracture for a uniaxial non-uniform tensile stress field

Weibull defined the "risk of rupture" B from the probability of non-fracture of a rod of volume V composed of unit rod volumes subjected to a stress σ uniformly distributed over the sectional area. Under the above assumption that the probabilities of non-failure in unit rod volumes are independent:

$$\mathrm{Ln}\,(1-P) = V\,\mathrm{Ln}\,(1 - P_{ov}) \qquad [2.13]$$

$$B = -\,\mathrm{Ln}\,(1 - P) \qquad [2.14]$$

where P = probability of fracture of a rod of volume V, P_{ov} = probability of fracture for a length of rod corresponding to unit volume.

For a small volume element dV, the risk of rupture is:

$$dB = -\,\mathrm{Ln}\,(1 - P_{ov})\,dV = n(\sigma)\,dV \qquad [2.15]$$

$n(\sigma)$ corresponds to the stress function $\phi(\sigma)$ introduced above in equation [2.5]. As discussed previously, Weibull considered that it is the material function expressing the strength properties of the material. For an isotropic brittle material, it is independent of the position of the volume element dV and the direction of the stress.

For a given stress distribution in a body:

$$B = \int_v n(\sigma) dV \qquad [2.16]$$

The corresponding failure probability is derived from equation [2.14]. The reference volume V_0 was introduced according to the above comment on scale factor dimension.

$$P(\sigma, V) = 1 - \exp\left[-\frac{1}{V_o}\int_v \left(\frac{\sigma}{\sigma_o}\right)^m dV\right] \qquad [2.17]$$

where $\sigma = \sigma(x, y, z)$ describes the stress state in the body with volume V. x, y and z are the coordinates.

2.4. Probability of fracture from the surface of specimens

In certain circumstances, probability of fracture from the surface must be determined, either because heterogeneities are also located in the surface, or because fracture initiates mainly from heterogeneities located in the surface of considered body.

For a uniform uniaxial stress state:

$$P(\sigma, A) = 1 - \exp\left[-\frac{S}{A_o}\left(\frac{\sigma}{\sigma_o}\right)^m\right] \qquad [2.18]$$

For a non-uniform uniaxial stress-state:

$$P(\sigma, A) = 1 - \exp\left[-\frac{1}{A_o}\int_S \left(\frac{\sigma}{\sigma_o}\right)^m dA\right] \qquad [2.19]$$

$\sigma = \sigma(x, y)$ is the stress in the surface. A is the area of the stressed surface. A_o is the reference area. $A_o = 1\ m^2$ in accordance with the international system. m and σ_o are the statistical parameters.

2.5. Weibull multiaxial analysis

In the preceding formulae, σ is a pure tensile stress. The formulae for uniform polydimensional stresses were derived considering that those normal stresses which are included within a spatial angle contribute to the risk of rapture. It was assumed that fracture is caused by normal stress components only, which implies that the compressive stresses and the shear stresses have no influence on the risk of rupture. The normal stress component at a given point depends on principal stresses and directions as shown in Figure 2.2:

$$\sigma_n = \cos^2\Phi\ (\sigma_1 \cos^2\Psi + \sigma_2 \sin^2\Psi) + \sigma_3 \sin^2\Phi \qquad [2.20]$$

where σ_n is the normal stress component, σ_1, σ_2 and σ_3 are the principal stresses and Φ and Ψ are the angles as shown in Figure 2.2.

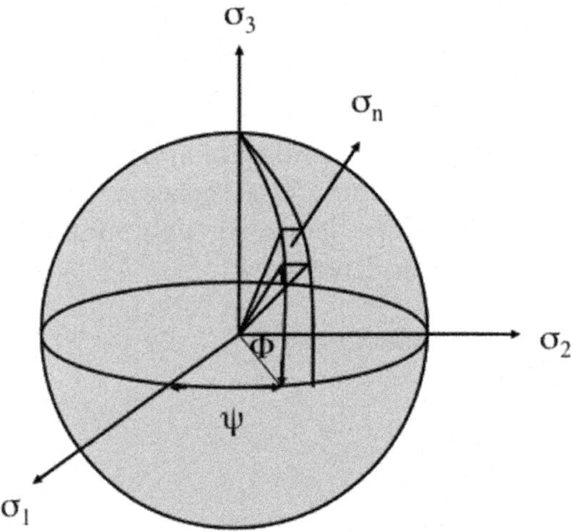

Figure 2.2. *Unit sphere illustrating relations between tensile normal stresses σ_n on planes and principal stresses σ_1, σ_2 and σ_3*

The normal stresses which are included within the small spatial angle do = cosΦ dΦdΨ, contribute to the risk of rupture an amount $n(\sigma_n)$ do. According to equation [2.16], the risk of rupture at given point p is expressed as:

$$B_p = \int_A n(\sigma_n) \cos\Phi d\Phi d\Psi \qquad [2.21]$$

If $n(\sigma_n) = k\,(\sigma_n)^m$, the probability of rupture is:

$$P = 1 - \exp\left[-\frac{k}{V_o}\int_V\int_A \sigma_n^m \cos\Phi d\Phi d\psi dV\right] \qquad [2.22]$$

The integral is over the hemispherical surface of the unit sphere surrounding point p, over those directions in which tensile stress occurs. This integration domain is designated by A in equations [2.21] and [2.22].

For the case of plane stress ($\sigma_3 = 0$), the risk of rupture is:

$$B = 2k\int_V \int_0^{\pi/2} \int_0^{\psi_o} \cos^{2m+1}\Phi(\sigma_1 \cos^2\Phi + \sigma_2 \sin^2\Phi)^m \, d\Phi d\Psi dV \qquad [2.23]$$

If σ_1 and σ_2 are tensile stresses, the limit of integration in equation [2.23] is $\psi_o = \frac{\pi}{2}$. If σ_2 is negative, $\psi_o = \sqrt{\frac{\sigma_1}{-\sigma_2}}$ T_g.

In his 1939 paper, Weibull derived from these formulae the risk of rupture for two-dimensional and three-dimensional tensile stresses for various values of m.

This approach to the treatment of uniform multiaxial stresss is based on definition of risk of rupture within a phenomenological approach. The risk of rupture has some functional relation to the applied tensile stress, clarified using a power-law function of the tensile normal stress.

The statistical distributions of flaw strengths, and the flaw response to stresses, are not specified. The intrinsic value of the material function $n(\sigma_n)$ expressing the strength properties of the material, and its relationship with $n(\sigma)$ pertinent to a uniaxial tension raises some questions. It is assumed that $n(\sigma)$ and η (σ_n) are described by the same function. It will be shown in subsequent chapters of this book on the multiaxial elemental strength model that the flaw strength density function is dependent on the stress-state, and that it can be related to the reference one defined on uniaxial conditions.

The risk of fracture depends on the power-law distribution function of the normal stress component. In other words, the material functions $n(\sigma a)$ in multiaxial stress state) or $n(\sigma_n)$ (in multiaxial stress state) are implicitly based on the following criterion for fracture: (or σ_n) $> \sigma_R$

where σ_R is the isotropic material resistance to uniaxial traction which is considered to be independent of the physical nature and orientation of flaws. The material function thus refers to the fraction of the flaw population, including those flaws the fracture of which obeys the above failure criterion, i.e. those flaws with orientation such that extension will be perpendicular to the normal stress direction. These comments will be discussed in Chapter 4 which establishes a more physically oriented probabilistic approach to fracture based on the concept of multiaxial elemental strength. Furthermore, it will be shown that this approach allows these issues to be clarified, as it provides physical meaning to the empirical material function.

In Chapter 4, an equivalent stress σ_{eq} can be defined using an appropriate multiaxial fracture criterion. This equivalent stress can combine normal, compressive and shearing stress components. Then, its statistical distribution provides failure probability within the context of the multiaxial elemental strength probabilistic model. It will be shown that the Weibull model is a particular case owing to the assumptions it is based upon.

Alternatively, a few authors directly introduced an equivalent stress into equation [2.22]:

$$P = 1 - \exp\left[-\frac{k}{V_0}\int_V \int_A \sigma_{eq}^m \cos\Phi d\Phi d\Psi dV\right] \qquad [2.24]$$

This approach is also discussed in Chapter 4.

2.6. Multiaxial approach based on the principle of independent action of stresses

Barnett and Freudenthal proposed a simple approach to the treatment of uniform multiaxial stresses within the context of the Weibull theory. This approach has been often adopted. It is assumed that the tensile principal stresses are independent and do not interact. According to the weakest link principle, the non-failure probability of the body is taken as the product of the individual survival probabilities for each tensile principal stress:

$$1 - P = (1 - P_1)(1 - P_2)(1 - P_3) \qquad [2.25]$$

where P_1, P_2 and P_3 are the fracture probabilities corresponding, respectively, to the principal stresses $\sigma_1 > \sigma_2 > \sigma_3$. They are given by the following equations for non-uniform tensile uniaxial stress state:

$$P_1 = 1 - \exp\left[-\frac{1}{V_o}\int_V \left(\frac{\sigma_1}{\sigma_0}\right)^m dV\right] \qquad [2.26]$$

$$P_2 = 1 - \exp\left[-\frac{1}{V_o}\int_V \left(\frac{\sigma_2}{\sigma_0}\right)^m dV\right] \qquad [2.27]$$

$$P_3 = 1 - \exp\left[-\frac{1}{V_o}\int_V \left(\frac{\sigma_3}{\sigma_0}\right)^m dV\right] \qquad [2.28]$$

Introducing these expressions into equation [2.24], it follows that:

$$P = 1 - \exp\left[-\frac{1}{V_o}\int_V \left(\frac{\sigma_1^m + \sigma_2^m + \sigma_3^m}{\sigma_0^m}\right) dV \qquad [2.29]$$

It appears that the probability of fracture takes form similar to equation [2.17] and that the resultant stress in equation [2.28] is different from the normal stress given by equation [2.20], which expresses a combination of principal stresses. The principle of independent action of principal stresses is a rough approximation. This technique has been found to predict failure probabilities well below the experimental one in the literature [RUF 84]. It is inherently non-conservative due to the fact that only the contributions to failure from each individual local principal stress component are considered, neglecting the effects of combined local principal stresses.

2.7. Summary on the Weibull statistical model

The Weibull fracture theory is in essence phenomenological at macroscopic level. It implies the presence of sources of fracture located at any point. In other words, it focuses on the symptoms instead of the causes of fracture: it does not recognize the flaws as being unique entities operated by stresses; the physical processes producing fracture are not addressed. It predicts that the brittle fracture strength is a function of stressed volume (or surface area), and stress-state.

The approach involves a material function expressing the strength properties of the material. This function cannot be considered as a material characteristic since the response of fracture-inducing flaws depends on various extrinsic factors such as flaw orientation and stress-state. The power law function proposed for $n(\sigma)$ is advantageous for manipulation of equations, and it is valid for a variety of materials under uniaxial tension. Furthermore, it concerns only a fraction of the population of flaws, i.e. those with preferred orientation to applied stresses. The contribution of compressive normal stresses is discarded, although they can combine to tensile normal components. Several authors in the literature have pointed out that any statistical analysis of failure must account for flaws oriented for growth under mixed-mode fracture conditions.

The Weibull distribution presents, however, significant advantages:

– it is used worldwide to model life data. It is flexible and it fits a wide range of data, including normal distributed data;

– the 2-parameter Weibull cumulative distribution has an explicit equation, involving two parameters that can be estimated easily from experimental failure data. This is an additional reason why the Weibull approach is the most widely used, primarily for routine analysis of failure data;

– it provides a satisfactory description of the statistical distribution of strength data for a wide variety of materials (including bulk specimens and fibers) when tested under simple loading conditions such as traction and bending.

However, as pointed out by several investigators, it presents some important limitations that question its ability to predict failure of large components having complex geometries and subjected to multiaxial stress-states. It presents descriptive virtue but limited predictive capacity. The resistance to fracture was found to have been overestimated for many cases. Examples are given in Chapter 9. This limitation has been pointed out within the context of the extreme value theory [BOU 83] for which the parameters of the law must be estimated from the extreme values of a finite set of data. Sound predictions cannot be obtained if the law is poorly known and if the physical processes governing the phenomenon are not described explicitly and exhaustively.

More physically oriented fracture theories that overcome the limitations of the Weibull model are discussed in subsequent chapters.

3

Statistical-Probabilistic Theories Based on Flaw Size Density

3.1. Introduction

In those theories which consider flaws as physical entities, the expression of failure probability is derived from flaw density function. This function is the probability density function (pdf), or density, of a random variable characteristic of flaw severity. The following characteristics are generally used:

– either a critical dimension like length for cracks;

– or flaw strength.

Crack size is related to stress by fracture mechanics expressions involving toughness given by K_{IC} (critical stress intensity factor), G_C (strain energy release rate) or J_C (J integral). The pdf describes the relative likelihood for the flaw severity variable to take a given value. The probability of failure is the probability of this variable falling within a particular range. Hence, it is given by the integral of this variable density over that range.

In this chapter, flaw severity is characterized by flaw size. Flaws are implicitly considered tantamount to cracks. The stress-state is uniaxial and uniform. Approaches based on flaw strength will be discussed in Chapter 4.

3.2. Failure probability

The flaw density function f(a) (Figure 3.1) is the density of probability P that X = a, in case of a continuous random variable:

$$f(a) = P(X = a) \quad [3.1]$$

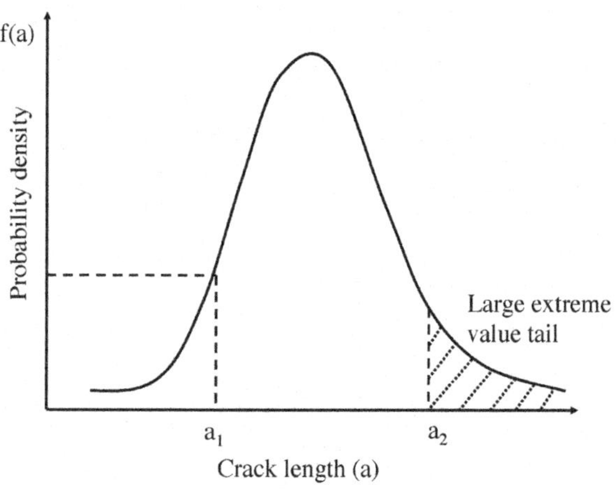

Figure 3.1. *Flaw size probability density function f(a)*

X is a random variable, and a is crack size. The cumulative distribution function, i.e. the probability that X ≤ a, is given by the integral of this density in the [0, a] range (Figure 3.2):

$$F(a) = P(X \leq a) = \int_{o}^{a} f(a)\, da \quad [3.2]$$

The probability of existence of flaws with size ≥ a is derived from equation [3.2]:

$$P(X \geq a) = \int_{a}^{+\infty} f(a)\, da \quad [3.3]$$

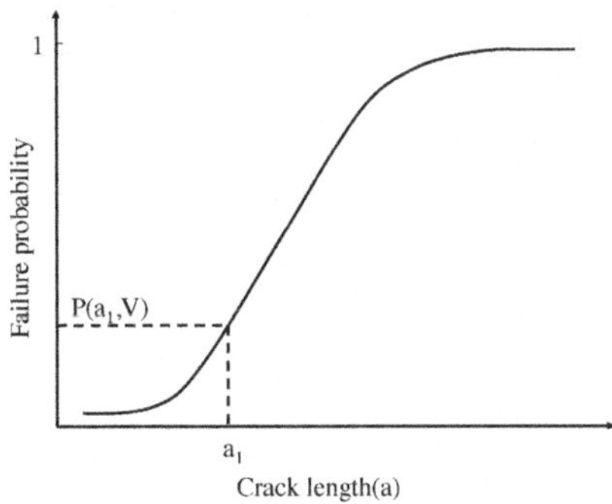

Figure 3.2. *Flaw size cumulative distribution function given by expression [3.2]*

In a volume element[1] V_o, the probabilty of fracture from a flaw with critical size a_1 is given by the probability of presence of a flaw with size $\geq a_1$.

$$P(X \geq a_1) = V_o \int_{a_1}^{+\infty} f(a)\, da = P(a_1, V_o) \qquad [3.4]$$

For a body having volume V = N V_o, the failure probability is obtained from the product of non-failure probabilities of volume elements, according to the weakest link principle:

$$1 - P(a_1, V) = (1 - P(a_1, V_o))^N = (1 - V_o \int_{a_1}^{+\infty} f(a)\, da)^N \qquad [3.5]$$

Introducing $V_o = V/N$ in equation [3.5], we obtain:

$$1 - P(a_1, V) = (1 - \frac{V}{N} \int_{a_1}^{+\infty} f(a)\, da)^N \qquad [3.6]$$

Expression [3.6] is the $\left(1-\frac{x}{N}\right)^N$ function which tends toward e^{-x} when $N \rightarrow +\infty$, i.e. when the number of elements is large. For a body

1 The volume element V_o must not be mistaken for the reference volume V_0 defined in Chapter 2.

with volume V, failure probability from a flaw with size a_1 is expressed as:

$$P(a_1,V) = (1 - \exp[-V \int_{a_1}^{+\infty} f(a)\,da] \qquad [3.7]$$

Application of equation [3.7] for failure probability computation requires an explicit formula for f(a) and the relation between flaw size and stress. The stress-state determines flaw severity, and particularly the value of a_1.

3.3. Expressions for flaw size density and distribution

The expression of f(a) can be selected arbitrarily among available classical formulae, or established from measurement of flaw sizes using microscopy or any other pertinent means of material analysis. The f(a) expressions pertain to two groups:

– on the one hand those which address the whole distribution of flaws (Figure 3.1);

– on the other hand, those which consider only the large value extreme (Figure 3.1).

3.3.1. *Description of complete flaw population*

There are many probability distributions that can be selected. Hereafter examples of distributions that have been used in the literature are given.

– First, the Gaussian distribution:

$$f_N(a) = \frac{1}{S_D\sqrt{2\pi}} \exp - \left[\frac{a-\mu}{S\sqrt{2}}\right]^2 \qquad [3.8]$$

where μ is the mean value of a, S_D is the standard deviation.

In [FOR 12], almost all the flaws present on the surface of a glass filament, including the most severe ones and those that would not cause failure, were detected using atomic force microscopy. Gaussian distribution satisfactorily fitted the distribution of flaw sizes that was obtained when considering all the flaws.

– Second, the beta distribution:

The pdf of the beta distribution has been introduced in models of fracture to express the distribution of critical flaw sizes. Calard *et al.* used the following expression (Figure 3.3):

$$f_{\alpha\beta}(x) = \frac{1}{B(\alpha,\beta)} x^{\alpha-1}(1-x)^{\beta-1} \text{ for } 0 \leq x \leq 1 \qquad [3.9]$$

where B (α, β) is the beta distribution:

$$B(\alpha,\beta) = \frac{\Gamma(\alpha)\Gamma(\beta)}{\Gamma(\alpha+\beta)} \qquad [3.10]$$

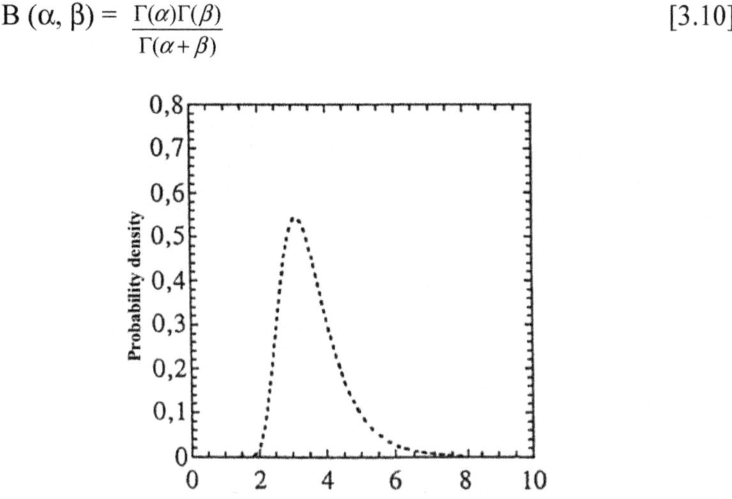

Figure 3.3. *Flaw size probability density function derived from the beta function for a SiC matrix (equation [3.8])*

Γ(.) is the gamma function, α and β are positive shape parameters that control the shape of the distribution. x was set to a/a_w, where a is the current flaw size and a_w is the largest mean flaw size.

Hild *et al.* used the following expression that takes into account the size of the most critical flaw a_M.

$$f(a) = \frac{a_M^{-1-\alpha-\beta}}{B_{\alpha\beta}} a^{\alpha} (a_M - a)^{\beta} \text{ for } 0 \leq a \leq a_M \quad [3.11]$$

$B\alpha\beta = B(\alpha+1, \beta+1)$, α and β are the shape parameters.

– Third, Poloniecki's empirical distribution:

Experimental results produced by Poloniecki are fitted closely by the following expression for f(a):

$$f(a) = \frac{c^{n-1}}{(n-2)!} a^{-n} e^{-c/a} \quad [3.12]$$

where c is a scaling parameter and n is the rate at which the density tends to 0 with increasing flaw size (Figure 3.4). This expression for flaw size distribution is the basis of the probabilistic failure model that Jayatilaka developed for a uniaxial tensile loading case. This model is discussed in section 3.5.2.

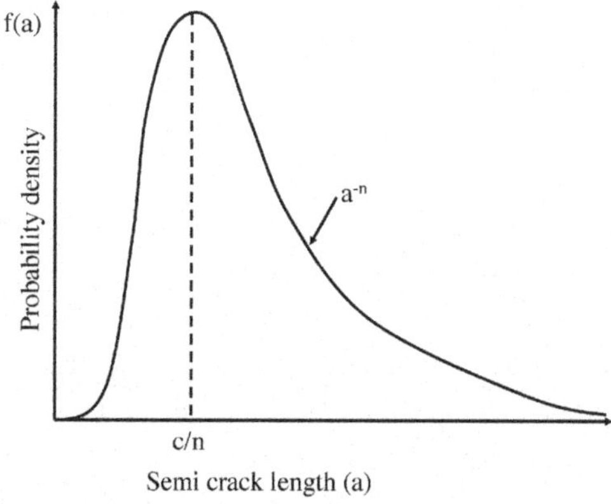

Figure 3.4. *Probability density function obtained by Poloniecki (expression [3.10])*

3.3.2. *Statistical distribution functions for extreme values*

Fracture of a brittle material is caused by one of its largest flaws, so that fracture strength is controlled by the large flaw size extreme of the flaw population. The tail of the distribution function can be described using distributions of extreme values.

Extreme value theory is the theory of modeling and measuring events which occur with very small probability. In the general literature of mathematics and statistics, the Weibull distribution is cited as the type III standard distribution function. Type I is the Gumbel distribution function (equation [3.13]), and type II is that of Fréchet (equation [3.14]).

$$F(x) = \exp\,[-\,\exp\,(-q\,[x - q^*])] \qquad [3.13]$$

$$F(x) = \exp\,[-\,(\frac{q^*}{x})^{\alpha}\,] \qquad [3.14]$$

where q and q* are the parameters.

Extreme value distributions depend on the rate of decrease of the original population toward the asymptote or the upper bound. When the rate of decrease is smooth (Cauchy-type distributions), a power-law expression can be used for the tail of the density function:

$$\lim f\,(a) = (\frac{q}{a})^{\lambda} \quad a \rightarrow +\infty \qquad [3.15]$$

The Pareto distribution may also warrant consideration, since it is bounded at the lower extreme value, which may be of interest when the flaws have a minimum size (a_{min}):

$$F(a) = 1 - (a_{min}/a)^k \qquad [3.16]$$

In [FOR 12], it was shown that the normal distribution expression fitted quite well the Weibull plot of experimental failure-derived flaw sizes for glass fibers (Figure 3.5). The Pareto distribution was also found to fit the Weibull plot of flaw sizes well (Figure 3.5). However, it is worth pointing out that for one type of glass fibers, the standard deviation could not be defined for the Pareto distribution owing to the exponent value k < 2 (k = 1.8). In [HU 85], a statistical theory of fracture in a single-phase and a two-phase material with flaws whose size follows a Paretto distribution was developed.

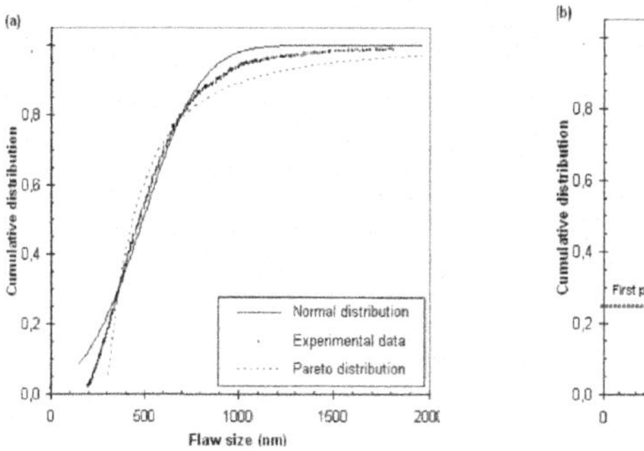

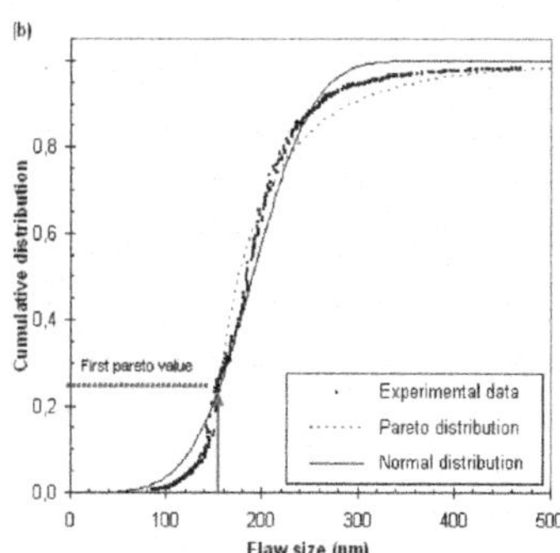

Figure 3.5. *Comparison of Normal distribution and Pareto distribution with Weibull plot of flaw sizes derived from tensile tests on two types of glass fibers*

3.4. Introduction of stress state

In linear fracture mechanics, size of crack is related to operating stress through the stress intensity factor:

$$K_I = \sigma Y \sqrt{a} \qquad [3.17]$$

where Y is a dimensionless geometry factor, K_I is the mode I stress intensity factor and σ is the stress operating on crack. At fracture under σ, critical crack size is given by the following expression derived from [3.17]:

$$\frac{K_{IC}^2}{\sigma^2 Y^2} = a(\sigma) \qquad [3.18]$$

Equation [3.18] is generally used to derive the expression for flaw strength density f(σ) from f(a). Crack length a represents the size of cracks perpendicular to stress direction. When cracks are inclined, appropriate expressions for a(σ) are required. De Jayatilaka developed a model using the stress necessary to fracture an inclined crack.

3.5. Models

Statistical approaches to brittle fracture based on flaw size distribution were not the subject of a large amount of papers. In those developments based on the beta distribution, a closed-form expression of fracture probability as a function of stress-state was not produced. Implicit equations were applied to computations of fracture probability under various loading conditions on a brittle solid [HIL 92], or to description of fragmentation in the matrix of a SiC/SiC fiber reinforced composite [CAL 96].

Two examples of developments of closed-form expressions of failure probability based on flaw size distribution are given in the following.

3.5.1. *Power-law flaw size density*

Introducing expression [3.15] in [3.7] and integrating gives:

$$P(a_1, V) = 1 - \exp\left[-V \frac{q^\lambda}{\lambda - 1}\left(\frac{1}{a_1}\right)^{\lambda-1}\right] \qquad [3.19]$$

The failure probability can be expressed in terms of stress by applying relation [3.18]:

$$P(\sigma, V) = 1 - \exp\left[-V \frac{q^\lambda Y^{2\lambda-2}}{(\lambda-1)K_{IC}^{2\lambda-2}} \sigma^{2\lambda-2}\right] \qquad [3.20]$$

This expression takes the Weibull form for a uniaxial stress-state (equation [3.22]), for the following expressions of m, σ_0:

$$\begin{cases} m = 2\lambda - 2 \\ \sigma_o = \dfrac{Y}{K_{IC}}\left[\dfrac{2}{m} q^{\frac{m+2}{2}}\right]^{1/m} \end{cases} \qquad [3.21]$$

$$P(\sigma, V) = 1 - \exp\left[-V\left(\frac{\sigma}{\sigma_o}\right)^m\right] \qquad [3.22]$$

Similarity with the Weibull expression was expected since the Weibull analysis uses power-law for the function n(σ) (Chapter 2), which represents the low extreme value in the distribution of fracture stresses, equivalent to the large extreme value in the distribution of fracture-inducing flaw sizes through the linear fracture mechanics relation [3.18].

3.5.2. *The De Jayatilaka–Trustrum approach*

De Jayatilaka and Trustrum developed a general expression for the failure probability of a brittle material by using the flaw size density function identified by Poloniecki (equation [3.12]) and the stress necessary to fracture an inclined crack. On the basis of strain energy concepts, the crack size and stress are related to the critical stress intensity factor and the crack angle. For small values of σ and large number of cracks, the implicit equation of failure probability takes the closed-form:

$$P = 1 - \exp\left[-N\,\frac{c^{n-1}}{n!}\left(\frac{\Pi\sigma^2}{K_{IC}^2}\right)^{n-1}\right] \qquad [3.23]$$

The above expression [3.23] may be written as:

$$P = 1 - \exp\left[-N\,k\,\sigma^{m}\right] \qquad [3.24]$$

where:

$$m = 2n - 2 \quad k = \frac{c^{m/2}}{(\frac{m+2}{2})!}\,\frac{\Pi^{m/2}}{K_{IC}^m} \qquad [3.25]$$

The number of cracks N is proportional to the stressed volume. Expression [3.23] is similar to the Weibull equation. Again, this similarity results from the extreme value hypothesis for the flaw size and associated failure stress, and from the power-law form for the flaw size density function. Thus, for large crack size values, the Poloniecki function becomes:

$$\frac{c^{n-1}}{(n-2)!}a^{-n} = (\frac{q}{a})^n \qquad [3.26]$$

where $q = \left[\frac{c^{n-1}}{(n-2)!}\right]^{1/n}$ [3.27]

3.6. Limits of the flaw size density-based approaches

Compared to the Wiebull theory, the statistical approaches to fracture based on flaw size density function are more fundamental, physically oriented since they are more explicitly concerned with the flaws that cause fracture. But, they face difficulties and present limitations. They were not very developed.

The detection of flaws for the determination of the flaw size density function is a major limitation. There are non-destructive methods based on image analysis available, and much progress has been made in this area. However, the non-destructive methods are unable to detect all the pertinent heterogeneities as well as the inclined cracks, and to distinguish the potential fracture-inducing flaws. A second issue is the definition of size for those flaws with a complex shape. Destructive methods can also be used. However, they cannot be considered as a reasonable alternative. They require fractographic analysis of a large number of test specimens. Furthermore, when possible, it is more convenient to extract fracture stresses directly from the fracture tests.

Crack size may be a satisfactory characteristic for those cracks perpendicular to stress direction in a uniform uniaxial stress state. Determination of the sizes of inclined cracks is not straightforward. It can be expected that results are obtained with a big uncertainty.

The developments presented above clarify the physical meaning of the material function (or stress function) introduced by Weibull in the risk of rupture. They show the relationship between the flaw distribution and the Weibull expression of failure probability. This enables us to highlight the significance of the form of the underlying flaw size density function.

Closed-form expressions for failure probability were obtained essentially for a uniaxial uniform stress-state and cracks loaded in mode I. More appropriate criteria are required to address dependence of flaw severity on stress states. Furthermore, the analysis of failure in multiaxial stress states is not straightforward. Numerical approaches can be used to overcome certain difficulties. Models based on the beta function are able to consider a wider population of flaws as requested by the analysis of matrix fragmentation in ceramic composites.

When the extreme of the flaw size density is considered, and when it is expressed as a power-law, expression of failure probability is reduced to the Weibull equation. This approach may be sufficient for certain materials, but it is not necessarily the most pertinent. There is no objective reason that alternative flaw size density functions would not be required depending on the material considered.

4

Statistical-Probabilistic Theories Based on Flaw Strength Density

4.1. Introduction

In these statistical-probabilistic theories, the severity of flaws is characterized by a critical stress.

In the elemental strength approaches, the expression for failure probability is derived from a flaw strength density function (Figure 4.1). This function is the probability density function of random variable flaw strength, which is defined as the elemental flaw extension stress, i.e. the critical value of the tensile stress operating locally on volume element that contains the flaw (Figure 4.2). The flaws are assumed to be spatially random and independent. In most models, a uniaxial element strength is considered [ARG 59, ARG 66, MAT 76, EVA 76, EVA 78] then a multiaxial elemental strength developed. In this chapter, much emphasis is placed on the failure model based on multiaxial elemental strength (multiaxial elemental strengh model) [LAM 83, LAM 88b, LAM 90].

The probability density function describes the relative likelihood of the elemental strength variable taking a given value. The probability of failure is the probability of this variable falling within a particular range. Hence, it is given by the integral of this variable density over that range.

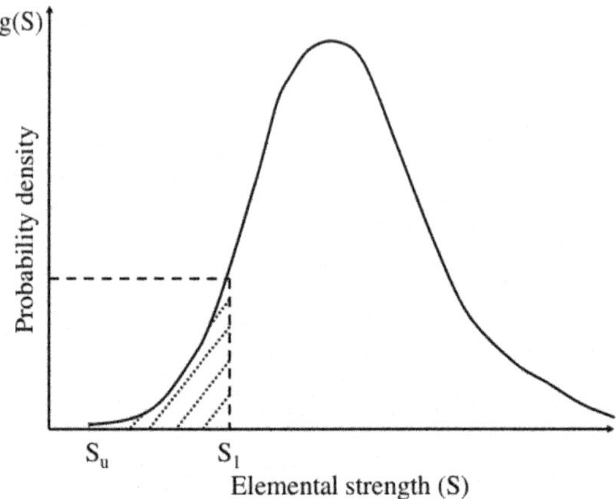

Figure 4.1. *Flaw strength density function g(S)*

Batdorf took a different route. He proposed a statistical theory in which the flaws are assumed to be cracks, and, therefore to have the directional sensitivity of cracks to the applied stresses. In the Batdorf's theory, the fracture characteristics of a material are completely determined by a function which represents the number of cracks/unit volume that will fracture when the tension normal to the plane of the crack exceeds the critical normal stress of a particular crack. Failure probability is given by the product of a probability of existence of favorable cracks and probability of crack orientation direction.

4.2. Basic equations of failure probability in the elemental strength approach

The theory is parallel to the one described in Chapter 3. It is detailed for clarity purposes. Another parallel theory could be obtained with an alternative flaw characteristic.

The flaw strength density function g(S) (Figure 4.1) is the density of probability P that X = S, in case of a continuous random variable:

$$g(S) = P(X = S) \qquad [4.1]$$

X is a random variable, and S is the elemental strength for a given flaw. The cumulative distribution function, i.e. the probability that $X \leq S$, is given by the integral of the density function in the [0, S] range (Figure 4.1):

$$F(S) = P(X \leq S) = \int_{o}^{S} g(S)\,dS \qquad [4.2]$$

The probability of existence of flaws more severe than a particular flaw is the probability that the associated elemental strengths are smaller than the elemental strength S_1 pertinent to the particular flaw (Figure 4.1). It is derived from equation [4.2]:

$$P(X \leq S_1) = \int_{o}^{S_1} g(S)\,dS \qquad [4.3]$$

Equation [4.3] gives the probability of fracture under stress $\sigma = S_1$. In volume element V_o[1], subjected to uniform tensile stress, the probability of fracture is given by the probability of presence of a flaw with elemental strength $\leq S_1$.

$$P(X \leq S_1) = V_o \int_{o}^{S_1} g(S)\,dS = P(S_1, V_o) \qquad [4.4]$$

For stressed volume $V = N\,V_o$, failure probability is obtained from the product of non-failure probabilities of volume elements, according to the weakest link principle:

$$1 - P(S_1,V) = (1 - P(S_1,V_o))^N = [1 - V_o \int_{o}^{S_1} g(S)\,dS]^N \qquad [4.5]$$

1 As indicated in Chapter 3, V_o should not be mistaken for V_0 that denotes the reference volume (Chapter 2).

As shown in Chapter 3, equation [4.5] tends toward the following exponential expression when the number of volume elements is large:

$$P(S_1,V) = 1 - \exp\left[-V \int_{o}^{S_1} g(S)\, dS\right] \qquad [4.6]$$

This equation applies when the stress state is uniaxial and uniform. Furthermore, it must be pointed out that the presentation of this theory is made general for scalar stresses. The direction of S_1 with respect to the directions of flaw extension and stress is not addressed. These issues are addressed in the models discussed in the following. In the first two models, the flaw strength is defined as stress perpendicular to crack plane. In the third model, it is the resultant of multiaxial stresses operating on flaw. In the first model, the stress state is uniaxial and uniform. In the other two models, it is non-uniform and polyaxial.

For uniaxial and non-uniform stress state, the probability of fracture is given by the following equation:

$$P = 1 - \exp\left[-\int dV \int_{o}^{S_1} g(S)\, dS\right] \qquad [4.7]$$

The case of multiaxial stresses is detailed in subsequent sections.

4.3. Elemental strength model for a uniform uniaxial stress state: Argon–McClintock development

Figure 4.1 shows an example of density function of elemental strengths. A few authors tried to determine g(S) from the load in many different kinds of tests [ARG 59, ARG 66 , MAT 76, EVA 83, EVA 88, EVA 90]. An example of determination of fracture probability under uniaxial tension is discussed. No assumption was made about the functional form of g(S). Furthermore, S was considered as a scalar, which implies that the orientation of S with respect to stress direction was not taken into account. It is thus assumed that the elemental strength is perpendicular to the flaw extension plane and parallel to the stress direction, as shown in Figure 4.2. The hatched portion of the density function (Figure 4.1) corresponds to the probability that flaw strength S is smaller than S_1. The elemental strength distribution exhibits a lower bound S_u that characterizes the most severe flaw pre-existing in the material. As a result, the function $g(S-S_u)$ is represented as a sum of terms that are calculated from the values of the function's derivative at a single point (Taylor series):

$$g(S - S_u) = B(S - S_u)^{m'} + \ldots \quad [4.8]$$

where B is a numerical coefficient depending on the *n*th derivative of the function evaluated at the point $(S-S_u)$, m' is a constant.

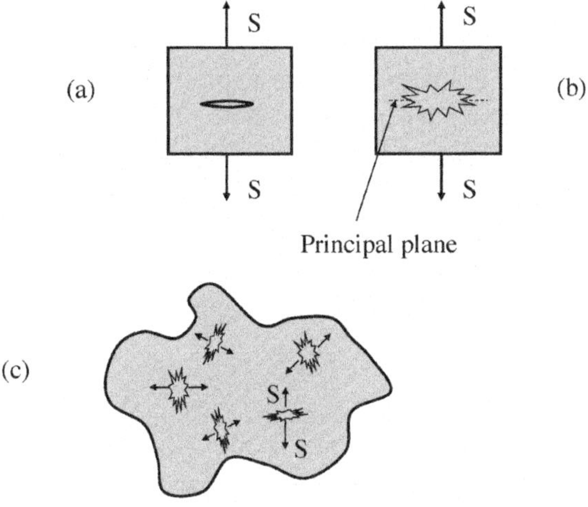

Figure 4.2. *The concept of uniaxial elemental stress for flaws and cracks: a) Uniaxial elemental stress for a crack b) Uniaxial elemental stress for a flaw c) Flaws and associated uniaxial elemental stresses in a material*

Failure probability for volume V subjected to a stress $\sigma = S_1$ is expressed as:

$$P(S_1, V) = 1 - \exp\left[-\frac{V}{V_o}\int_{S_u}^{S_1}\left(\frac{S - Su}{S^*}\right)^{m'} dS\right] \quad [4.9]$$

with $B = \dfrac{V_o}{(S^*)^{m'}}$

Integrating gives:

$$P(S_1, V) = 1 - \exp\left[-\frac{V}{(S^*)^{m'}(m'+1)}(S_1 - S_u)^{m'+1}\right] \qquad [4.10]$$

Replacing appropriate terms with the following constants m and S', equation [4.10] becomes:

$$m = m' + 1;\ (S')^{m} = (S^*)^{m'}(m'+1) \qquad [4.11]$$

$$P(S_1, V) = 1 - \exp\left[-V\left(\frac{S_1 - S_u}{S'}\right)^{m}\right] \qquad [4.12]$$

Equation [4.12] is identical to the expression proposed by Weibull for a uniform uniaxial tensile stress state. This result provides some justification for the power law form of the stress function $n(\sigma)$ (also referred to as $\phi(\sigma)$ (equation [2.7])), which corresponds to the low extreme of the elemental strength distribution for those flaws with a particular orientation with the stress state. In this particular case, $g(S)$ and $\phi(\sigma)$ are related by the following expression:

$$g(S) = (m' + 1)\,\frac{\phi(S)}{S_1 - S_u} \qquad [4.13]$$

This similarity is dependent on the basic assumptions made. A more rigorous treatment considering all the flaws including those which inclined will lead to different equations of failure probability.

4.4. The Batdorf model

4.4.1. *The model*

Treatment by Batdorf seems to have been inspired by statistical thermodynamics. Like molecules in the kinetic theory of gases, flaws have a random location and a random orientation.

Microscopic flaws are assumed to be flat cracks. Flaw strength is characterized by critical stress σ_{CR} which is given by the specific stress that causes crack extension in mode I. Consequently, it is perpendicular to the crack plane. Flaws are assumed to be independent.

Two necessary conditions must be met at fracture: first, cracks with critical stress $\sigma_{CR} > 0$ must be present in the material, and the local stress σ_e must exceed the critical stress σ_{CR} for at least one flaw. The local stress σ_e is the stress normal to the crack plane, or the effective stress corresponding to the fracture criterion selected. Note that there is no condition on the direction of σ_e that is defined as a scalar combining stress components. The probability of fracture is the product of the probabilities of occurrence of the above conditions.

$$P_f(\Delta V, \sigma_e) = P_1 \,.\, P_2 \qquad [4.14]$$

where P_1 is the probability of existence in volume ΔV of a crack having a critical stress between σ_{CR} and $\sigma_{CR} + d\sigma_{CR}$ and P_2 denotes the probability that σ_{CR} will be oriented in a direction such that $\sigma_e \geq \sigma_{CR}$. For a uniform uniaxial stress state in ΔV

$$P_1 = \Delta V \, \frac{dN(\sigma_{CR})}{d\sigma_{CR}} \, d\sigma_{CR} \qquad [4.15]$$

where $N(\sigma_{CR})$ represents the number of cracks/unit volume with a critical stress $\leq \sigma_{CR}$.

$$P_2 = \frac{\Omega(\sigma, \sigma_{CR})}{4\pi} \qquad [4.16]$$

where Ω represents the solid angle containing the normals to all orientations for which $\sigma_{CR} < \sigma_e$, σ signifies the applied state of stress.

Failure probability for a volume element subjected to a uniform stress state ($\sigma_e = \sigma$) is expressed as:

$$P_f(\Delta V, \sigma) = \Delta V \int_o^\infty \frac{\Omega}{4\pi} \frac{dN(\sigma_{CR})}{d\sigma_{CR}} \, d\sigma_{CR} \qquad [4.17]$$

Failure probability for a volume V, made up of elements ΔV is given, as above, by the limit of the function in expression [4.19] when the number of volume elements takes large values. According to the weakest link concept, the non-failure probability is the product of non-failure probabilities of volume elements:

$$P_S(V, \sigma) = 1 - P_f(V, \sigma) = [1 - P_f(\Delta V, \sigma)]^{V/\Delta V} \quad [4.18]$$

$$= [1 - \Delta V \int_o^{\infty} \frac{\Omega}{4\pi} \frac{dN(\sigma_{CR})}{d\sigma_{CR}} d\sigma_{CR}]^{V/\Delta V} \quad [4.19]$$

When: $\frac{V}{\Delta V} \rightarrow \infty$

$$P_S(V, \sigma) = \exp - [V \int_o^{\infty} \frac{\Omega}{4\pi} \frac{dN(\sigma_{CR})}{d\sigma_{CR}} d\sigma_{CR}] \quad [4.20]$$

Failure probability for a uniform stress state is:

$$P_f(V, \sigma) = 1 - \exp [-V \int_o^{\infty} \frac{\Omega}{4\pi} \frac{dN(\sigma_{CR})}{d\sigma_{CR}} d\sigma_{CR}] \quad [4.21]$$

For a non-uniform uniaxial stress state, failure probability is expressed as:

$$P_f(V, \sigma) = 1 - \exp [-\int_V dV \int_o^{\infty} \frac{\Omega}{4\pi} \frac{dN}{d\sigma_{CR}} d\sigma_{CR}] \quad [4.22]$$

For calculation of failure probability for a material of volume V, expressions for the distribution function $N(\sigma_{CR})$ and the stress state are required. Batdorf first proposed that the function N is represented by a Taylor series, assuming that there is some lower limit to the critical stress σ_u.

$$N(\sigma_{CR}) = \sum_{j=0}^{\infty} b_j \left(\frac{\sigma_{CR}}{\sigma_u} - 1\right)^j \quad \sigma_{CR} \geq \sigma_u \quad [4.23]$$

$$N(\sigma_{CR}) = 0 \quad \sigma_{CR} < \sigma_u$$

Equation [4.23] represents the entire distribution of flaw strengths. When examining the case in which fracture is dictated by the most severe flaws at the low strength extreme, it is assumed that the distribution function can be approximated by a single term in the power series. It can be a bounded or unbounded power law:

$$N(\sigma_{CR}) = k_B \left(\frac{\sigma_{CR}}{\sigma_u} - 1\right)$$

$$N(\sigma_{CR}) = k_B\, \sigma_{CR}^{r} \qquad [4.24]$$

The solid angle Ω can be obtained by integrating $d\Omega$ over the range in which $\sigma_e \geq \sigma_{CR}$:

$$\Omega = 2 \int_{\beta=0}^{2\pi} \int_{\alpha=0}^{\pi} H(\sigma_e, \sigma_{CR}) \sin\alpha \; d\alpha\, d\beta \qquad [4.25]$$

where α and β are the azimuthal angles with respect to the principal stresses, and $H(\sigma_e, \sigma_{CR})$ is the Heaviside operator:

$$H(\sigma_e, \sigma_{CR}) = 1 \text{ when } \sigma_e \geq \sigma_{CR}$$

$$H(\sigma_e, \sigma_{CR}) = 0 \text{ when } \sigma_e < \sigma_{CR}$$

Analytical expressions can be found for Ω when the stress state can be expressed by simple analytical functions, like in simple tension and equal biaxial tension. In general, Ω must be determined numerically.

The function Ω depends on the effective stress σ_e through the fracture criterion selected, as well as on the stress state. When $\sigma_e = \sigma_n$, the stress normal to crack plane (σ_n) is related to principal stresses using the closed-form expression given in Chapter 2 (equation [2.20]). Although failure condition compares scalar quantities, it is necessary that σ_n and σ_{CR} are colinear vectors.

Batdorf proposed the following shear-dependent expressions for σ_e for two crack shapes (Griffith cracks (equations [4.26] and [4.28]) and penny-shaped cracks (equations [4.26] and [4.28]) using the maximum tensile stress (equations [4.26] and [4.27]) and strain-energy release rate considerations (equations [4.28] and [4.29]):

$$\sigma_e = 0.5\,[\sigma_n + \sqrt{\sigma_n^2 + \tau^2}\,] \qquad [4.26]$$

$$\sigma_e = 0.5\,[\sigma_n + \sqrt{\sigma_n^2 + \tau^2/(1-0,5\nu)^2}\,] \qquad [4.27]$$

$$\sigma_e = \sqrt{\sigma_n^2 + \tau^2} \qquad [4.28]$$

$$\sigma_e = \sqrt{\sigma_n^2 + \tau^2/(1-0,5\nu)^2} \qquad [4.29]$$

Normal stress components can be calculated at any point using the stress relations in the principal stress space, as indicated earlier for σ_n. Again the failure condition compares scalar quantities whatever the direction of stresses, which implies that σ_e is perpendicular to the crack plane. As a result, taking into account the shear component would increase the stress operating on the crack, which is not a proper reproduction of the real phenomenon of crack extension. The Batdorf model proposes an approach to shear sensitive coplanar extension of the crack. It may be appropriate for certain stress states. The function $N(\sigma_{CR})$ is considered independent of the stress state, as well as g(S). However, the response of flaws under local polyaxial stresses and the consequent dependence of $N(\sigma_{CR})$ and g(S) on polyaxial stress states must be taken into account. A more rigorous polyaxial treatment is proposed with the multiaxial elemental strength model. But, before, a few practical examples illustrate the calculation of fracture probability using the Batdorf model.

4.4.2. *Examples of determination of failure probability using the Batdorf model: unixial, equibiaxial and equitriaxial tension*

For simple tension, the solid angle consists of two circular cones as shown in Figure 4.3:

$$\Omega = 2\int_{o}^{\sigma_{CR}} 2\pi \sin\theta \, d\theta = 4\pi[1 - \cos\theta_{CR}] \qquad [4.30]$$

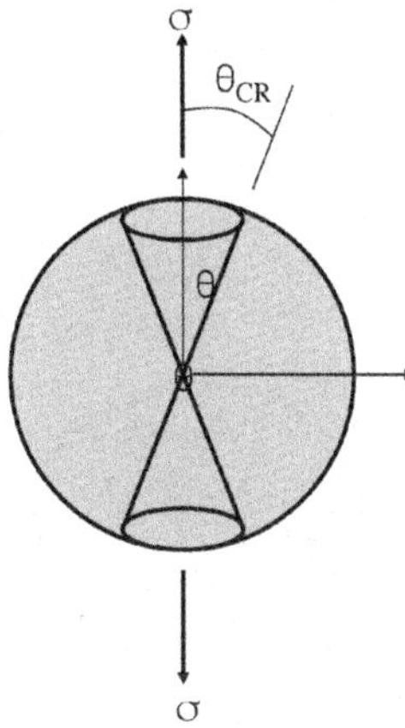

Figure 4.3. *Uniaxial tension: solid angles within which the crack normal must lie for fracture to occur*

where θ is the angle between the maximum principal stress axis and the normal to the crack. θ_{CR} delimits the portion of space where $\sigma_e \geq \sigma_{CR}$.

The normal stress component is related to applied stress by:

$$\sigma_n = \sigma \cos^2 \theta \qquad [4.31]$$

In the case when $\sigma_e = \sigma_n$, $\sigma_{CR} = \sigma \cos^2 \theta_{CR}$. Substituting $\cos \theta_{CR}$ in equation [4.30], it comes:

$$\Omega = 4\,\pi\,\left(1 - \sqrt{\frac{\sigma_{CR}}{\sigma}}\right) \qquad [4.32]$$

Using a similar method gives the following expressions for Ω, for the particular cases of *equibiaxial tension (Figure 4.4) and equitriaxial tension*:

equibiaxial tension:

$$\Omega = 4\,\pi\,\sqrt{1 - \frac{\sigma_{CR}}{\sigma}} \qquad [4.33]$$

equitriaxial tension:

$$\Omega = 4\,\pi \qquad [4.34]$$

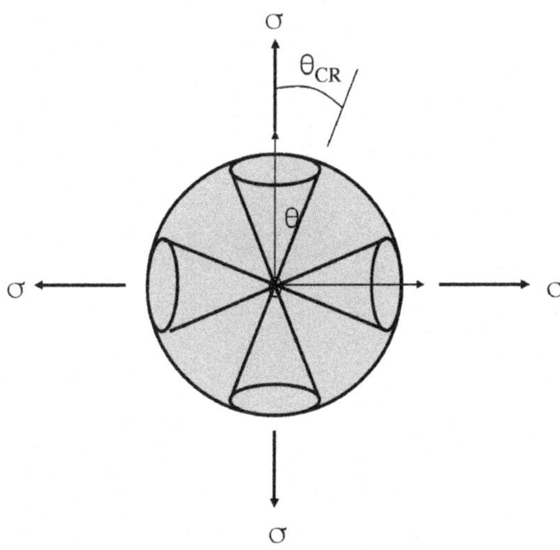

Figure 4.4. *Equibiaxial tension: solid angles within which the crack normal must lie for fracture to occur*

For the bounded distribution function $N(\sigma_{CR}) = k_B\,(\frac{\sigma_{CR}}{\sigma_u} - 1)^r$, the failure probability in uniaxial tension is expressed as:

$$P = 1 - \exp\,[-\,V\,k_B\,\frac{1}{2(r+1)}\,(\frac{\sigma}{\sigma_u} - 1)^{r+1}] \qquad [4.35]$$

For the unbounded distribution function $N(\sigma_{CR})$, the following expressions of failure probability are obtained for the particular tensile loading cases:

uniaxial tension:

$$P = 1 - \exp\,[-\,k_B\,V\,\frac{\sigma^m}{(2m+1)}] \qquad [4.36]$$

equibiaxial tension:

$$P = 1 - \exp\left[-k_B V \frac{\sigma^m m!}{(m+0,5)(m-0,5)...1,5}\right] \quad [4.37]$$

equitriaxial tension:

$$P = 1 - \exp[-k_B V\sigma^m] \quad [4.38]$$

4.4.3. *Discussion: comparison with the Weibull model*

The form of equations of failure probability proposed by the Batdorf theory is comparable to that of Weibull theory. This results from the underlying power law for crack density in the Batdorf model, and for the material function in the Weibull theory. Comparison of equation [4.35] obtained for a bounded distribution function to the corresponding Weibull equation [2.11] shows that both give the same failure probability for the following values of constants:

$$r + 1 = m$$

$$b_r \frac{1}{2(r+1)} = \left(\frac{\sigma_u}{\sigma_o}\right)^m \quad [4.39]$$

Then, comparing equation [4.36] for an unbounded distribution function to the corresponding Weibull equation (derived from [2.11] for $\sigma_u = 0$), we obtain:

$$k_B = \frac{(2m+1)}{\sigma_o^m} \quad [4.40]$$

Estimates of statistical parameters from the same set of strength data are equivalent according to equations [4.39] and [4.40]. Then, this equivalence is dependent on loading case, so that it can be expected that both models will provide different failure predictions. An example that was reported by Shetty is presented in Chapter 9. Table 9.2 shows that the relations between statistical parameters vary with loading case. It appears in Figures 9.2 and 9.3 that the Batdorf model gave strength predictions that were closer to the experimental results. The discrepancy between both models is enhanced when a fracture criterion different from the implicit normal stress criterion is introduced.

4.5. The multiaxial elemental strength model [LAM 83]

Since the flaws have any orientation to the applied stresses, they are operated locally by a multiaxial stress state. The elemental strength cannot be defined as the stress perpendicular to crack plane or to the direction of flaw extension. It must correspond to the combined action of normal and shear stress components.

The flaw strength density function g(S) is not a material intrinsic characteristic. It is dependent on stress state, and the relation with the uniaxial flaw strength density function must be established. Failure probability is then derived from equation [4.7].

4.5.1. *The multiaxial elemental strength*

The multiaxial elemental stress is defined as the stress (σ_E) equivalent to the local stress-state that operates on the flaw or the crack. The stress-state comprises a normal stress component perpendicular to flaw principal plane (σ_n), and a shear stress component parallel to this plane (τ). The principal plane coincides with the plane of flaw extension under uniaxial tension ($\tau = 0$) (Figure 4.5).

The equivalent elemental strength is defined through the strain energy release rate. The strain energy release rate depends on direction at notch tip. It takes maximum value in the direction of propagation. A flaw becomes a crack as soon as it starts propagating. Initial length corresponds to flaw size in the principal plane (Figure 4.5):

$$G_{max} = \frac{(1+\upsilon)(1+X)}{4E}\left[K_I^4 + 6K_I^2K_{II}^2 + K_{II}^4\right]^{1/2} \qquad [4.41]$$

with X = 3 – 4 ν for plane strain condition, and X = (3 – ν)/(1 + ν) for plane stress conditions.

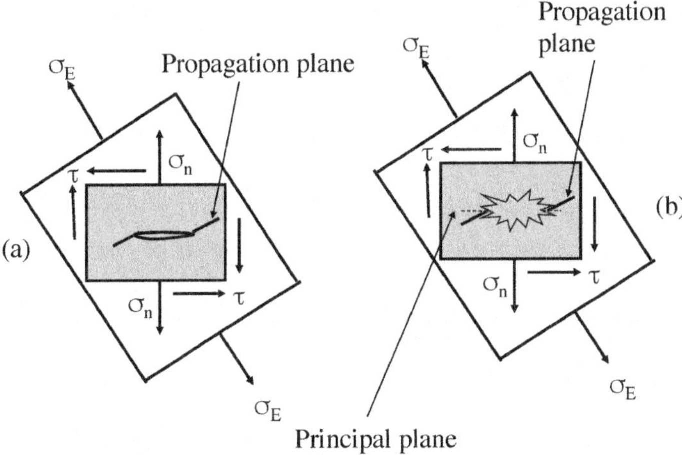

Figure 4.5. *The concept of multiaxial elemental stress for flaws and cracks a) For a crack b) For a flaw*

E is Young's modulus, ν is Poisson's coefficient, K_I and K_{II} are mode I and mode II stress intensity factors.

The direction of crack extension is given by the angle to the principal plane (γ_0). Its formula given by equation [4.42] is not necessary in the model:

$$\gamma_0 = \text{arctg}\left[-\frac{2K_I K_{II}}{K_I^2 + K_{II}^2}\right] \qquad [4.42]$$

Substituting K_I and K_{II} given by the following expressions into equation [4.41], we obtain for the equivalent elemental stress σ_E:

$$K_I = \sigma Y\sqrt{a}\,,\ K_{II} = \tau Y\sqrt{a}$$

$$\sigma_E = \left[\sigma_n^4 + 6\tau^2\sigma_n^2 + \tau^4\right]^{1/4} \qquad [4.43]$$

In short, σ_E appears as an equivalent stress combining local stress components σ_n and τ, which induces the same value of G_{max} as σ_n and

τ do. The resistance of flaws is determined by the critical value of elemental stress (S_E), attained when G_{max} takes the critical value G_c:

$$S_E = \left[\frac{4EG_c}{(1+\upsilon)(1+X)} \right]^{1/2} Y\sqrt{a} \qquad [4.44]$$

4.5.2. *The flaw density function (volume analysis)*

The flaw density function is now the distribution of multiaxial elemental strengths $g_E(S_E)$. The number of flaws in a volume element must be the same whatever the loading case, with $g_E(S_E)$ for multiaxial elemental strength or g(S) for uniaxial elemental strength perpendicular to flaw principal plane. For a unit volume element sphere (Figure 4.6), the number of flaws with a resistance between S and S+dS is given by the following formula:

$$g(S)\, dS\, 4\,\pi = 2 \int_o^{\pi/2} \int_o^{\pi} g_E(S_E) dS_E \cos\varnothing d\varnothing d\psi \qquad [4.45]$$

Angles $\varnothing$ and ψ indicate the direction normal to flaw or crack plane (Figure 4.6).

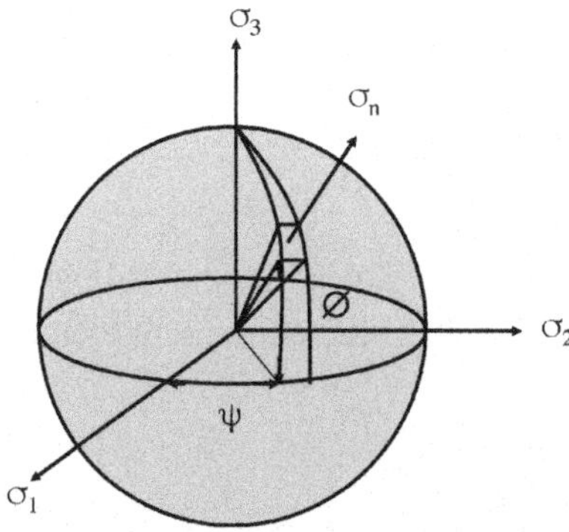

Figure 4.6. *Unit sphere illustrating the relations between local stresses (σ_n, τ) on planes and principal stresses σ_1, σ_2 and σ_3*

Like in the approaches discussed earlier, an explicit closed-form expression is required for $g_E(S_E)$, in order to establish a formula for failure probability. The power law was found to be satisfactory for flaw strength distribution g(S) as shown in Chapter 9. It allowed sound failure predictions for ceramics under various loading conditions. However, the estimated statistical parameters generally exhibit some variation. This issue has not been solved properly, despite much effort by many researchers. Alternative flaw strength distributions can be identified, as indicated in Chapter 3. The normal distribution should be a natural solution. It is considered the most prominent probability distribution in statistics. It indicates the probability of occurrence of a characteristic in a population of infinite size. Then, certain distributions can be approximated by the normal distribution when the sample size is large (for example, the binomial distribution, the Poisson distribution, the χ-squared distribution and Student's t-distribution). It is reported that this trend is also observed with the Weibull distribution when the shape parameter $3<m<4$. Normal distribution is not used in fracture statistics. It has been assumed by a few researchers for the distribution of strengths in the locality of flaws [PEI 26, EPS 48, LU 02, RMI 12].

A power law corresponding to the first term of a Taylor series was selected for $g_E(S_E)$:

$$g_E(S_E) = m\, S_E^{m-1} S_o^{-m} \qquad [4.46]$$

m and S_0 are the statistical parameters that can be estimated from experimental data as shown in Chapter 7. Equation [4.46] corresponds to the low strength extreme in a Weibull distribution (Chapter 10).

Substituting $g_E(S_E)$ into equation [4.45], we obtain for g(S):

$$g(S)\, dS = \frac{1}{2\pi} \int_o^{\frac{\pi}{2}} \int_o^{2\pi} m S_E^{m-1} S_o^{-m} dS_E \cos\varnothing\, d\varnothing\, d\psi \qquad [4.47]$$

4.5.3. *Determination of local stress components*

Firstly, place the stress state operating on the solid that is induced by the boundary conditions is known. It can be related to applied loads either by analytical equations for a few simple classic cases, or by numerical analysis for complex loading cases. The stress components in the vicinity of flaws are defined in the space of principal stresses as shown in Figure 4.6. The stresses σ_n and τ that operate in planes defined by $\varnothing$ and ψ are given by the following equations:

$$\sigma_n = \sigma_1 \cos^2\varnothing \cos^2\psi + \sigma_2 \cos^2\varnothing \sin s^2\psi + \sigma_3 \sin^2\varnothing$$

$$\tau^2 = \sigma_1^2 \cos^2\varnothing \cos^2\psi + \sigma_2^2 \cos^2\varnothing \sin s^2\psi + \sigma_3^2 \sin^2\varnothing - \sigma_n^2 \qquad [4.48]$$

where σ_1, σ_2 and σ_3 are the principal stresses ; $\sigma_1 > \sigma_2 > \sigma_3$. σ_2 and σ_3 can be negative (compressive components) when σ_1 and $\sigma_E > 0$. The resulting failure probability is > 0. Otherwise, failure probability = 0.

Substituting σ_n and τ into equation [4.43], the corresponding equivalent strength is:

$$S_E = \sigma_1 \left[\left(\frac{\sigma_n}{\sigma_1}\right)^4 + 6\left(\frac{\sigma_n}{\sigma_1}\right)^2\left(\frac{\tau}{\sigma_1}\right)^2 + \left(\frac{\tau}{\sigma_1}\right)^4 \right]^{1/4} \qquad [4.49]$$

$$S_E = \sigma_1 f_v \left(\varnothing, \psi, \frac{\sigma_2}{\sigma_1}, \frac{\sigma_3}{\sigma_1}\right) \qquad [4.50]$$

when $\sigma_1 < 0$, $f_v \left(\varnothing, \psi, \frac{\sigma_2}{\sigma_1}, \frac{\sigma_3}{\sigma_1}\right) = 0$

After introduction of the expression of S_E in [4.47], we obtain

$$g(S)\, dS = \frac{1}{2\pi} \int_0^{\pi/2} \int_0^{2\pi} m \frac{\sigma_1^{m-1}}{S_o^m} f_v^m \, d\sigma_1 \cos\varnothing \, d\varnothing \, d\psi \qquad [4.51]$$

The corresponding equation of failure probability when fracture initiates from internal flaws is obtained as:

$$P_v = 1 - \exp\left[-\int_v \frac{dV}{V_o}\int_o^S \frac{1}{2\pi}\int_0^{\pi/2}\int_o^{2\pi} m\frac{\sigma_1^{m-1}}{S_o^m} f_v^m d\sigma_1 \cos\varnothing d\varnothing d\psi\right] \quad [4.52]$$

It can be rewritten as:

$$P_v = 1 - \exp\left[-\frac{1}{V_o}\int_v dV\int_o^S m\frac{\sigma_1^{m-1}}{S_o^m} f_v^m d\sigma_1 \frac{1}{2\pi}\int_o^{\pi/2}\int_o^{2\pi} f_v^m \cos\varnothing d\varnothing d\psi\right] \quad [4.53]$$

The expression of failure probability becomes:

$$P_v = 1 - \exp\left[-\frac{1}{V_o}\int_V dV\left(\frac{\sigma_1}{S_o}\right)^m I_v\left(m, \frac{\sigma_2}{\sigma_1}, \frac{\sigma_3}{\sigma_1}\right)\right] \quad [4.54]$$

with

$$I_v\left(m, \frac{\sigma_2}{\sigma_1}, \frac{\sigma_3}{\sigma_1}\right) = \frac{1}{2\pi}\int_o^{\pi/2}\int_o^{2\pi} f_v^m \cos\varnothing d\varnothing d\psi \quad [4.55]$$

4.5.4. *Probability of failure from surface flaws*

Determination of the expression of failure probability involves the same steps. The expressions of flaw strength density function and local stresses in the vicinity of flaws are different. The main equations are given hereafter:

$$g(S)\, dS = \frac{1}{2\pi}\int_o^{2\pi} g_E(S_E) dS_E d\psi \quad [4.56]$$

$$\sigma_n = \sigma_1 \cos^2\psi + \sigma_2 \sin^2\psi \quad [4.57]$$

$$\tau = (\sigma_1 - \sigma_2) \cos\psi \sin\psi$$

where σ_1 and σ_2 are the principal stresses ($\sigma_1 > \sigma_2$) (Figure 4.7):

$$S_E = \sigma_1\ f_s\left(\psi, \frac{\sigma_2}{\sigma_1}\right)$$

when $\sigma_1 < 0$, $f_s(\psi, \frac{\sigma_2}{\sigma_1}) = 0$

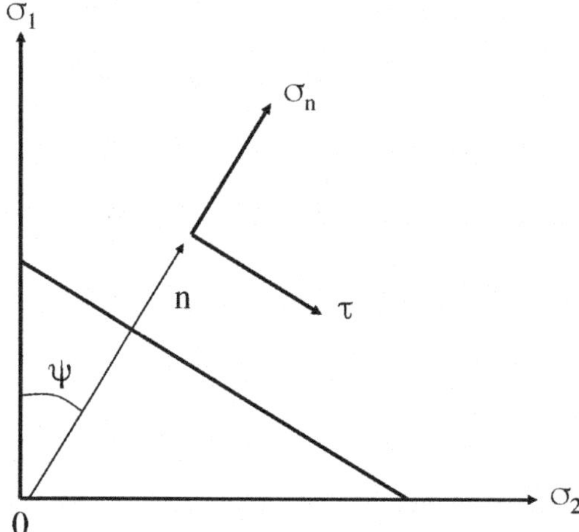

Figure 4.7. *Elementary Cauchy tetrahedron illustrating the relations between local stresses (σ_n, τ) on planes and principal stresses σ_1, σ_2*

The corresponding expression of failure probability is:

$$P_S = 1 - \exp\left[-\frac{1}{A_o}\int_A dA\left(\frac{\sigma_1}{S_o}\right)^m I_S(m, \frac{\sigma_2}{\sigma_1})\right] \qquad [4.58]$$

with

$$I_S(m, \frac{\sigma_2}{\sigma_1}) = \frac{1}{2\pi}\int_o^{2\pi} f_s^m(m, \frac{\sigma_2}{\sigma_1})d\psi \qquad [4.59]$$

4.5.5. *Determination of functions $I_V(...)$ and $I_S(...)$*

Closed-form solutions of loading functions $I_V(...)$ and $I_S(...)$ can be established for those stress states that are described by simple analytic expressions such that the expressions of local stress components as functions of principal stresses can be introduced into equation [4.50].

– *Uniaxial tension – fracture from surface flaws*

Substituting the principal stress components (σ_1 = constant = σ_1, $\sigma_2 = 0$) into equation [4.57], the local stress components are obtained:

$$\sigma_n = \sigma_1 \cos^2\psi$$

$$\tau = \sigma_1 \cos\psi \sin\psi$$

Substituting σ_n and τ into equation [4.43] of σ_E, it comes:

$$f_s(\psi, 0) = [\cos^8\psi + 6\cos^6\psi \sin^2\psi + \cos^4\psi \sin^4\psi]^{1/4} \quad [4.60]$$

The corresponding expression of I_s (m, 0) is:

$$I_s(m, 0) = \frac{2}{\pi}\int_o^{\pi/2} \cos^m\psi[1 + 4\sin^2\psi\cos^2\psi]^{m/4} d\psi \quad [4.61]$$

– *Uniaxial tension – fracture from internal flaws*

The local stress components are derived from equation [4.48] for the following expressions of principal stresses: σ_1 = constant = σ_1, $\sigma_2 = \sigma_3 = 0$:

$$\sigma_n = \sigma_1 \cos^2\psi \cos^2\varnothing \;(\frac{\tau}{\sigma_1})^2 = \cos^2\varnothing \cos^2\psi - \cos^4\varnothing \cos^4\psi \quad [4.62]$$

Substituting σ_n and τ into equation [4.43] of σ_E, it comes:

$$f_v(\varnothing, \psi, 0, 0) = \cos\varnothing \cos\psi[\,1 + 5\cos^2\varnothing \cos^2\psi\,(1 - \cos^2\varnothing \cos^2\psi)]^{1/4} \quad [4.63]$$

The corresponding expression of I_v (m, 0,0) is obtained from [4.55]:

$$I_v(m, 0,0) = \frac{2}{\pi}\int_o^{\pi/2}\int_o^{\pi/2} \cos^{m+1}\varnothing \cos^m\psi[1 + 5\cos^2\varnothing\cos^2\psi(1 - \cos^2\varnothing\cos^2\psi)]^{m/4} d\varnothing d\psi] \quad [4.64]$$

There is an alternate method of calculating I_v (m, 0, 0) which takes advantage of symmetry about the principal direction (σ_1 axis). The defect orientation is defined by angle θ (Figure 4.8). The expressions for the local stress components are:

$$\sigma_n = \sigma_1 \cos^2\theta \qquad [4.65]$$

$$\tau = \sigma_1 \cos\theta \sin\theta$$

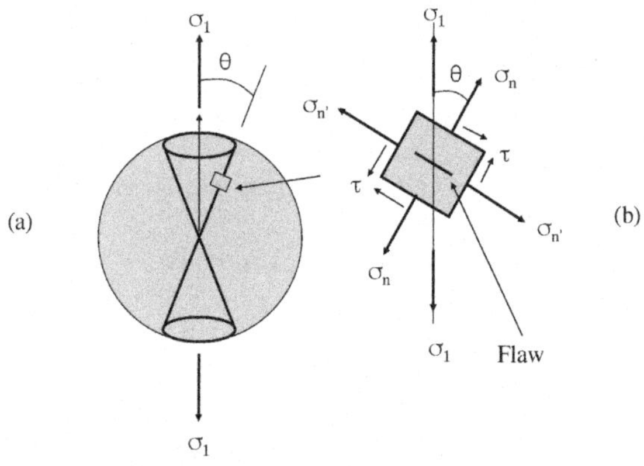

Figure 4.8. *Uniaxial tension: diagram showing the local stresses operating on a volume element containing a flaw and inclined by angle θ to principal stress direction σ_1.*

Substituting σ_n and τ into equation [4.43] of σ_E, we obtain for f_v (θ, 0, 0):

$$f_v(\theta, 0, 0) = \cos\theta\,[1 + 4\sin^2\theta\cos^2\theta\,]^{1/4} \qquad [4.66]$$

Introducing the expression of f_v (θ, 0, 0) in [4.55], we obtain for I_v (m, 0, 0):

$$I_v(m, 0, 0) = \frac{1}{4\pi}\int f_v^m\,dV \qquad [4.67]$$

$$dV = 2 \text{ x } 2\,\pi \sin\theta\, d\,\theta$$

$$I_v(m, 0, 0) = \int_0^{\pi/2} \cos^m\theta \sin\theta\,[1 + 4\sin^2\theta\cos^2\theta]^{m/4}\, d\,\theta \qquad [4.68]$$

– *Equibiaxial tension – fracture from surface flaws*

The local stress components are derived from equation [4.57] for the following expressions of principal stresses: $\sigma_1 = \sigma_2$:

$$\sigma_n = \sigma_1$$

$$\tau = 0$$

Substituting σ_n and τ into equation [4.43] of σ_E, we obtain:

$$f_s(\psi, 1) = 1 \quad [4.69]$$

Substituting f_s into equation [4.59], we obtain for I_s (m, 1):

$$I_S(m, 1) = \frac{2}{\pi}\int_o^{\pi/2} d\psi = 1 \quad [4.70]$$

– *Equibiaxial tension – fracture from internal flaws*

The local stress components are derived from equation [4.48] for the following expressions of principal stresses: $\sigma_1 = \sigma_2$, $\sigma_3 = 0$:

$$\sigma_n = \sigma_1 \cos^2\varnothing$$

$$\tau = \sigma_1 \cos\varnothing \sin^2\varnothing \quad [4.71]$$

Substituting σ_n and τ into equation [4.43] of σ_E, we obtain:

$$f_s(\varnothing, 1, 0) = [1 + 4\cos^2\varnothing \sin^2\varnothing]^{1/4} \cos\varnothing \quad [4.72]$$

Substituting f_v ($\varnothing$, 1, 0) into equation [4.55], we obtain:

$$I_v(m, 1, 0) = \int_o^{\pi/2} \cos^{m+1}\varnothing\, [1 + 4\cos^2\varnothing \sin^2\varnothing]^{m/4}\, d\varnothing \quad [4.73]$$

– *Equitriaxial tension – fracture from internal flaws*

The local stress components are derived from equation [4.48] for the following expressions of principal stresses: $\sigma_1 = \sigma_2 = \sigma_3$:

$$\sigma_n = \sigma_1$$

$$\tau = 0 \quad [4.74]$$

Substituting σ_n and τ into equation [4.43] of σ_E, we obtain:

$$f_v(\varnothing, \psi, 1, 1) = 1 \qquad [4.75]$$

Substituting $f_v(\varnothing, 1, 0)$ into equation [4.55], we obtain:

$$I_S(m, 1, 1) = \frac{2}{\pi}\int_o^{\pi/2}\int_o^{\pi/2} \cos\varnothing d\varnothing d\psi = 1 \qquad [4.76]$$

– *Computations of* $I_S(\ldots)$ *and* $I_V(\ldots)$

Except the loading cases of equibiaxial tension (fracture from the surface flaws) and equitriaxial tension, the loading functions $I_S(\ldots)$ and $I_V(\ldots)$ depend on m. Integration is generally difficult except when m takes certain particular integer values for which an analytic solution exists. Figure 4.9 depicts the influence of m on $I_S(\ldots)$ and $I_V(\ldots)$ values computed for exemplary loading conditions. Table 4.1 gives the corresponding values of $I_S(\ldots)$ and $I_V(\ldots)$.

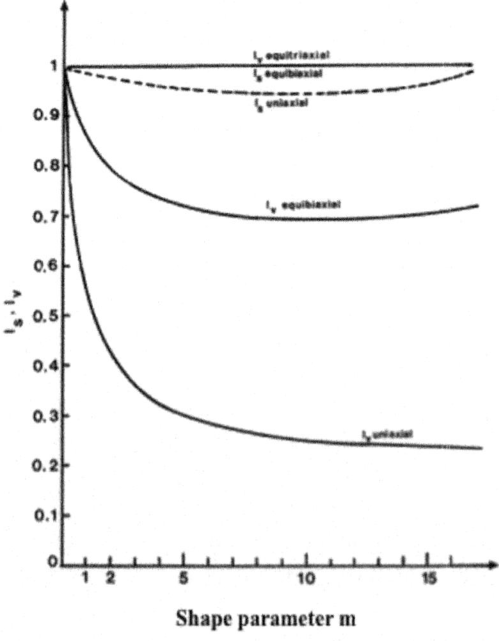

Figure 4.9. *Values of loading functions $I_s(\ldots)$ and $I_v(\ldots)$ versus parameter m for various loading cases*

	$I_s(m, \frac{\sigma_2}{\sigma_1})$	$I_V(m, \frac{\sigma_2}{\sigma_1}, \frac{\sigma_3}{\sigma_1})$	
m	$\frac{\sigma_2}{\sigma_1} = 0$	Uniaxial	equibiaxial
1	0.7048	0.5662	0.8645
2	0.608	0.431	0.7989
3	0.5598	0.3655	0.7607
4	0.5313	0.327	0.7365
5	0.5128	0.3017	0.7204
6	0.5003	0.2841	0.7096
7	0.4917	0.2712	0.7024
8	0.4858	0.2616	0.6979
9	0.4819	0.2543	0.6954
10	0.4795	0.2487	0.6945
11	0.4783	0.2443	0.6948
12	0.4781	0.241	0.6963
13	0.4786	0.2386	0.6987
14	0.4799	0.2368	0.7018
15	0.4817	0.2356	0.7057
16	0.484	0.2348	0.7102
17	0.4868	0.2345	0.7152
18	0.49	0.2346	0.7208
19	0.4936	0.235	0.7268
20	0.4975	0.2356	0.7333

Table 4.1. *Values of loading functions I_S and I_V for a range of m values, and for a few loading cases*

4.5.6. *Comparison with the Weibull model*

Differences can be noted from the comparison of equations of failure probability established in the multiaxial elemental strength model and in the Weibull model:

– first, the presence of loading functions I_S ( ) or I_V ( ) in the multiaxial elemental strength model that account for flaw orientation to stress directions. The values taken by these functions are smaller than 1 in many cases (Table 4.1 and Figure 4.9), which suggests that, other terms being equal, smaller probabilities should be predicted by the multiaxial elemental strength model;

– second, the scale factor S_0 that applies to the distribution function of elemental strengths in the multiaxial elemental strength model (flaw strength), whereas the Weibull scale factor (σ_o) refers to solid strength.

For further comparison of failure predictions, the values of respective scale factors are required. In Chapter 9, failure predictions for various loading cases are discussed.

The limitations of the multiaxial Weibull analysis were discussed in Chapter 2. In short, it is based on a material function $n(\sigma_n)$, which involves implicitly normal stress criterion for fracture. Instead, in the multiaxial elemental model, an equivalent stress is defined that is related to the current normal and shear stresses operating on the flaws.

Thus, when replacing σ_n with the local equivalent stress σ_E defined in the multiaxial elemental model, $\sigma_E = \sigma_1.f_V$, and assuming that the function $n(\sigma_n)$ is unchanged, we obtain from equation [2.22]:

$$P(\sigma_E) = 1 - \exp\left[-\frac{(2m+1)}{\Pi V_o}\int_V\left(\frac{\sigma_1}{\sigma_0}\right)^m dV\int_A f_v^m \cos\varnothing d\varnothing d\psi\right] \qquad [4.77]$$

Despite apparent formal similarity, there are the following important differences between equation [4.77] and equation [4.55] developed in the multiaxial elemental strength analysis: the constants, the scale factors and the integration domains. Obviously, the results of failure predictions will depend on the model used.

The multiaxial strength model provides a general equation for calculating failure probability for a wide variety of stress states. It is based on a flaw strength density function that accounts for flaw severity. Flaw severity depends on operating stresses. Being based on flaw strength, the multiaxial elemental strength model allowed evaluation of existing statistical-probabilistic approaches to brittle fracture.

By contrast, there is no continuity from the multiaxial to the uniaxial model in the Weibull analysis. Equation [2.17] cannot be derived from equation [2.22]. The material function $n(\sigma)$ has a different meaning in uniaxial and multiaxial conditions. In the multiaxial analysis, it refers to a local normal stress component (σ_n), while it refers to a uniaxial stress parallel to loading direction in the uniaxial analysis.

5

Effective Volume or Surface Area

5.1. Introduction

The notion of effective volume (respectively, effective surface area) is useful for the comparison of fracture resistance under different stress states, loading modes or volume sizes. The effective volume is a probabilistically equivalent volume of material subjected to a uniform tensile stress. It represents the volume of material for which failure probability under a specific uniform tensile stress is equal to that under the current stress state. The specific tensile stress is the maximum stress in the current stress state:

$$\sigma_{max} = \text{Max}\ [\sigma\ (x, y, z)]$$

where $\sigma\ (x, y, z)$ is the stress at point of coordinates (x, y, z).

The corresponding equation of failure probability is simple and unique whatever the stress state.

This chapter discusses the determination of effective volume for the Weibull and the multiaxial element strength models. Similar development can be made with the Batdorf approach.

5.2. The Weibull model: the effective volume for a uniaxial stress state

Equation [2.17] of failure probability can be rewritten as a function of maximum stress (σ_{max}) operating on solid:

$$P(\sigma, V) = 1 - \exp\left[-\frac{1}{V_o}\left(\frac{\sigma_{\max}}{\sigma_o}\right)^m \int_v \left(\frac{\sigma}{\sigma_{\max}}\right)^m dV\right] \quad [5.1]$$

The term $\int_v (\frac{\sigma}{\sigma_{\max}})^m dV$ having volume dimension represents the effective volume V_E. Substituting this term into equation [5.1], the failure probability expression becomes:

$$P(\sigma, V) = 1 - \exp\left[-(\frac{\sigma_{\max}}{\sigma_0})^m \frac{V_E}{V_0}\right] = P(\sigma_{\max}, V_E) \quad [5.2]$$

Equation [5.2] is the expression of failure probability for a solid having volume V_E, and operated by uniform stress of magnitude σ_{max}. V_E under σ_{max} is statistically equivalent to V under σ (x, y, z). The effective volume displays the following features:

– first, it has smaller size than the current volume under non-uniform stresses. In the presence of stress gradient, $\sigma / \sigma_{max} < 1$ over a certain volume Ω such that $\Omega \subset V$. Consequently, $\int_v (\frac{\sigma}{\sigma_{\max}})^m dV < \int_v dV$, and $V_E < V$;

– second, it has the size of current stressed volume when the stress state is uniform: when $\frac{\sigma}{\sigma_{\max}} = 1$, therefore $V_E = \int_v dV = V$. This implies the important feature that the maximun effective volume is obtained for a uniform tensile stress state. According to equation [5.2], the corresponding failure probability is the maximum attainable for given σ_{max} whatever the uniaxial stress state. Thus, it is shown that the tensile uniform stress state is the most severe.

An *effective surface area* can be defined similarly for fracture from surface flaws:

$$P = (\sigma, A) = 1 - \exp\left[-\frac{1}{A_0}\left(\frac{\sigma_{\max}}{\sigma_{0S}}\right)^{m_S} \int_A \left(\frac{\sigma}{\sigma_{\max}}\right)^{m_S} dA\right] \quad [5.3]$$

where σ_{max} is the maximum stress operating on the surface of solid. m_s and σ_{os} are the Weibull statistical parameters pertinent to the distribution of strengths.

$$A_E = \int_s \left(\frac{\sigma_{max}}{\sigma_{os}} \right)^{m_s} dA \qquad [5.4]$$

A_E represents the equivalent area of the outer surface of solid subjected to uniform uniaxial stress σ_{max}, such that the corresponding failure probability is identical to that of the current surface A subjected to stress state σ (x, y). As above with the effective volume, it is shown that the effective surface area for a solid subject to a non–uniform stress state is smaller than the outer surface area of the solid.

5.3. The multiaxial elemental strength model: the effective volume for a multiaxial stress state

As in the previous section, equation [4.54] of failure probability for a multiaxial stress state can be written as a function of maximum stress (σ_{max}) operating on the solid:

$$P = (\sigma, V) = 1 - \exp\left[-\frac{1}{V_o}\left(\frac{\sigma_{max}}{\sigma_{ov}}\right)^{m_v} \int_v \left(\frac{\sigma_1}{\sigma_{max}}\right)^{m_v} I_v(m_v, \frac{\sigma_2}{\sigma_1}, \frac{\sigma_3}{\sigma_1}) dV\right] \qquad [5.5]$$

where σ_1, σ_2 and σ_3 are the principal stresses such that $\sigma_1 > \sigma_2 > \sigma_3$. m_v and σ_{ov} are the statistical parameters pertinent to the flaw strength density function. In the following, we use σ_{ov} instead of S_0 for the scale factor in the flaw strength density function. However, it must be kept in mind that this scale factor is different from the Weibull one that refers to the statistical distribution of strengths of specimens.

Failure probability for the maximum stress σ_{max} operating uniformly and uniaxially over the effective volume is given by:

$$P(\sigma_{max}, V_E) = 1 - \exp\left[-\frac{1}{V_o}\left(\frac{\sigma_{max}}{\sigma_{oV}}\right)^{m_v} I_v(m_v 0{,}0) V_E \right. \qquad [5.6]$$

The expression of V_E is obtained by equating equations [5.5] and [5.6].

$$P(\sigma, V) = P(\sigma_{max}, V_E)$$

$$V_E = \frac{1}{I_v(m_v,0,0)}\int_v \left(\frac{\sigma_1}{\sigma_{max}}\right)^{m_v} I_v\left(m_v,\frac{\sigma_2}{\sigma_1},\frac{\sigma_3}{\sigma_1}\right)dV \qquad [5.7]$$

For current uniaxial stress state, it comes from equation [5.7] since I_v (m_v, 0,0) is a constant:

$$V_E = \frac{1}{I_v(m_v,0,0)}\int_v \left(\frac{\sigma_1}{\sigma_{max}}\right)^{m_v} I_v(m_v,0,0)dV \equiv \int_v \left(\frac{\sigma_1}{\sigma_{max}}\right)^{m_v} dV \qquad [5.8]$$

Note that this expression of effective volume is similar to that obtained above for the Weibull model for uniform stress state.

When the current multiaxial stress state is uniform $\sigma_1 = \sigma_{max}$, and $I_v(m_v,\frac{\sigma_2}{\sigma_1},\frac{\sigma_3}{\sigma_1})$ is a constant. It comes from equation [5.7] for the effective volume:

$$V_E = \frac{I_V\left(m_V,\frac{\sigma_2}{\sigma_1},\frac{\sigma_3}{\sigma_1}\right)}{I_V(m_V,0,0)}V \qquad [5.9]$$

Values of $I_v(.)$ are given in Table 4.1.

When the current multiaxial stress state is non-uniform, the computation of V_E using equation [5.7] requires a numerical method when closed-form equations for the stress state cannot be established, or when they are too complex.

The effective volume was defined with respect to a uniform uniaxial stress state for simplicity reasons. Uniaxial tensile stress seems to be a sound reference, since uniaxial tests can be carried out more easily than multiaxial tests. In subsequent sections, various effective volumes under various multiaxial stress states are defined.

When fracture occurs from surface-located flaws, the same approach yields analogous equations of equivalent surface area:

$$\mathrm{A_E} = \frac{1}{I_s(m_s,0)}\int_s \left(\frac{\sigma_1}{\sigma_{\max}}\right)^{m_s} I_s\left(m_s,\frac{\sigma_2}{\sigma_1}\right)dA \qquad [5.10]$$

As it was shown in the previous section, the effective surface area possesses the same properties as the effective volume: its maximum size is that of the stressed surface, which is obtained when a uniform tensile stress state prevails. The following general expressions of effective surface area are derived from equation [5.10] for simple stress states:

– for current uniaxial stress state, it comes from equation [5.10]:

$$\mathrm{A_E} = \frac{1}{I_s(m_s,0)}\int_s \left(\frac{\sigma_1}{\sigma_{\max}}\right)^{m_s} I_s(m_s,0)dA = \int_s \left(\frac{\sigma_1}{\sigma_{\max}}\right)^{m_s} dA \qquad [5.11]$$

As it was shown for effective volume, the expression of effective surface area is similar to that obtained for the Weibull model for a uniform stress state;

– for current multiaxial and uniform stress state ($\sigma_1 = \sigma_{max}$), it comes from equation [5.10]:

$$A_E = \frac{I_S\left(m_S,\dfrac{\sigma_2}{\sigma_1}\right)}{I_S(m_S,0)}A \qquad [5.12]$$

– for current multiaxial non–uniform stress state, $\mathrm{A_E}$ is computed using equation [5.10].

5.4. Analytic expressions for failure probability, effective volume or surface area (Weibull theory)

5.4.1. *Compression*

Compressive stresses are negative (σ (x, y, z) < 0), while the low strength limit is positive σ_u. Thus, σ (x, y, z) < 0 < σ_u. Consequently, from equation [2.9]:

$$P(V, \sigma) = 0$$

Local compressive stresses tend to close cracks and flaws. As a result, they cannot cause fracture. However, remote compressive stresses may generate tensile stresses locally. For instance, this is observed on the Brazilian test of diametral compression on a disk, for which tensile stresses operate on the central part of disk. It is worth pointing out that compressive loads should not be mistaken for local compressive stress components.

In a solid of which parts are subjected to compressive stresses, failure probability will be computed only in those parts operated on by tensile stresses. According to the weakest link concept, it comes:

$$1 - P(V) = (1 - P(\sigma_c, V_c))\,(1 - P(\sigma_t, V_t)) \qquad [5.13]$$

where V_c is the volume of material under compressive stress σ_c, V_t is the volume under tensile stress σ_t.

As discussed above $P(\sigma_c, V_c) = 0$. Consequently:

$$1 - P(V) = 1 - P(\sigma_t, V_t) \qquad [5.14]$$

$$P(V) = P(\sigma_t, V_t)$$

As expected, the effective volume for a solid subjected to a compressive stress state is $V_E = 0$.

5.4.2. *3-point bending*

Failure probability of the beam shown in Figure 5.1 is computed using equation [2.17], considering the volume under tensile stress. The upper half of the beam is subjected to compression, whereas the lower part is under a uniaxial tensile stress parallel to the z-axis:

For $x < 0$, $\sigma(x, y, z) < 0$

For $x > 0$, $\sigma(x, y, z) = \sigma_{max}\left(\frac{l-z}{l}\right)\left(\frac{x}{d}\right)$ [5.15]

with $\sigma_{max} = \frac{3Fl}{4bd^2}$.

Introducing expression of stress into equation [2.17], we obtain:

$$P(\sigma, V) = 1 - \exp\left[-\frac{1}{V_0}.\left(\frac{\sigma_{max}}{\sigma_0}\right)^m 2\int_o^l\int_o^d\left(\frac{l-z}{l}\right)^m\left(\frac{x}{d}\right)^m bdxdz\right] \quad [5.16]$$

Integrating the right-hand side, the exact expression of failure probability is obtained:

$$P(\sigma, V) = 1 - \exp\left[-\frac{1}{V_0}\left(\frac{\sigma_{max}}{\sigma_0}\right)^m \frac{V}{2(m+1)^2}\right] \quad [5.17]$$

where V is the volume of specimen: V= 4 bdl (see Figure 5.1 for beam dimensions).

When only the bottom face of the beam is considered (surface analysis), the stress is derived from equation [5.15] for x = d:

$$\sigma(x, y, z) = \sigma_{max}\left(\frac{l-z}{l}\right) \quad [5.18]$$

The corresponding failure probability is derived from equation [2.19]:

$$P(\sigma, A) = 1 - \exp\left[-\frac{1}{A_o}\left(\frac{\sigma_{max}}{\sigma_0}\right)^m \frac{A}{m+1}\right] \quad [5.19]$$

where A is the stressed surface area: A = 2 bl (Figure 5.1). The effective volume and the effective surface area are derived from equations [5.17] and [5.19]:

$$V_E = \frac{V}{2(m+1)^2} \quad [5.20]$$

$$A_E = \frac{A}{m+1} \quad [5.21]$$

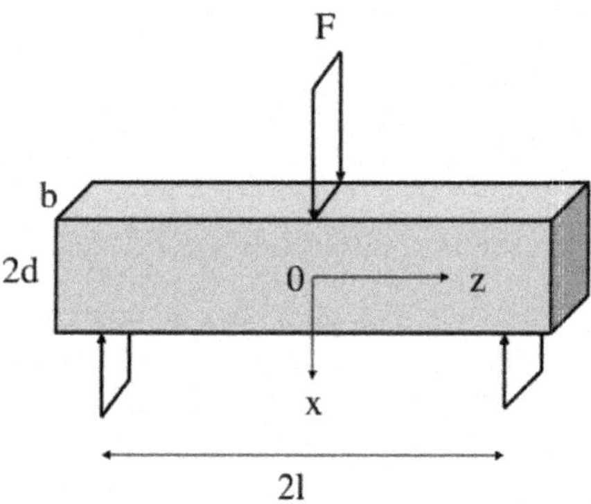

Figure 5.1. *The 3-point bending test configuration*

Depending on fracture origins, surface analysis considering the stresses on the flanks of the beam can also be performed.

5.4.3. *4-point bending*

As in 3-point bending, the upper half of the beam is operated on by compressive stresses (x < 0; Figure 5.2). Again, only the lower half of the beam under tensile stresses is considered for the computation of failure probability. The tensile stress is parallel to z-axis. It is uniform with respect to z-axis in the part of the beam located below the loads (Figure 5.2). The equations of stress state considered in a volume analysis are:

$$\sigma(x, y, z) = \sigma_{max}\left(\frac{x}{d}\right) \qquad 0 \leq z \leq l_1 \quad [5.22]$$

$$\sigma\,(x, y, z) = \sigma_{max}\,(\frac{l_2 - Z'}{l_2})\,(\frac{x}{d}) \quad 0 \leq z' \leq l_2 \qquad [5.23]$$

with $\sigma_{max} = \dfrac{3Fl_2}{4bd^2}$.

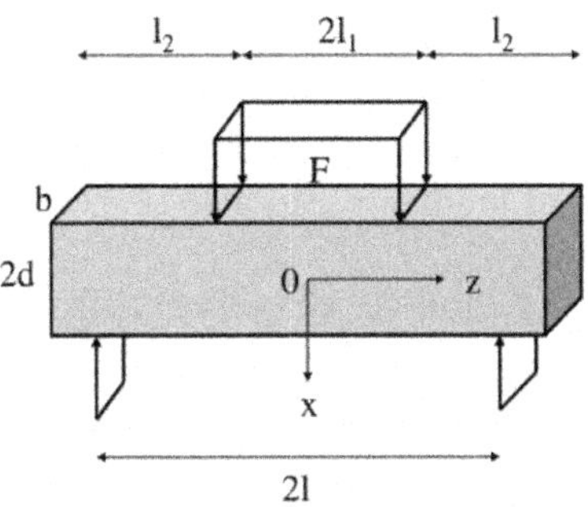

Figure 5.2. *The 4-point bending test configuration*

According to the weakest link principle, failure probability of the beam is obtained from the product of survival probabilities of the constitutive parts of the beam, as delineated by the locations of applied forces and supports. The corresponding expressions of failure probability are obtained from equation [2.17] for the above expressions of stresses. The failure probability of the beam is then:

$$P\,(\sigma, V) = 1 - \exp\,[-\frac{1}{V_o}(\frac{\sigma_{max}}{\sigma_0})^m \qquad [5.24]$$

$$[2\int_o^d (\frac{x}{d})^m bl_1 dx + 2\int_o^d\int_o^{l_2}\left(\frac{l_2 - z}{l_2}\right)^m (\frac{x}{d})^m bdxdz]$$

After integrating, it comes:

$$P\,(\sigma, V) = 1 - \exp\,[-\frac{1}{V_o}(\frac{\sigma_{max}}{\sigma_0})^m\left(\frac{2bdl_1}{m+1} + \frac{2bdl_2}{(m+1)^2}\right)] \qquad [5.25]$$

When only the bottom face of the beam is considered (surface analysis), the stresses are derived from equations [5.22] and [5.23] for x = d. Considering the constitutive portions of the bottom surface, and introducing the appropriate expressions of stresses into equation [2.19], the corresponding failure probability is:

$$P(\sigma, A) = 1 - \exp\left[-\frac{1}{A_o}\left(\frac{\sigma_{max}}{\sigma_0}\right)^m\left(2bl_1 + \frac{2bl_2}{m+1}\right)\right] \quad [5.26]$$

where A is the stressed surface area (A = 2bl; Figure 5.2).

Depending on fracture origins, a surface analysis considering the stresses in the flanks of the beam can also be performed.

The effective volume and the effective surface area (corresponding to the outer surface of the beam) are derived from equations [5.25] and [5.26]:

$$V_E = \frac{2bdl_1}{m+1} + \frac{2bdl_2}{(m+1)^2} \quad [5.27]$$

$$A_E = 2bl_1 + \frac{2bl_2}{m+1} \quad [5.28]$$

In the following, expressions of failure probability for fracture from the volume or the bottom surface of the beam are derived from equations [5.25] and [5.26], for typical cases corresponding to particular positions of applied forces:

$$l_1 = l_2 = \frac{l}{2}$$

$$P(\sigma, V) = 1 - \exp\left[-\frac{1}{V_o}\left(\frac{\sigma_{max}}{\sigma_0}\right)^m \frac{bdl(m+2)}{(m+1)^2}\right] \quad [5.29]$$

$$P(\sigma, A) = 1 - \exp\left[-\frac{1}{A_o}\left(\frac{\sigma_{max}}{\sigma_o}\right)^m \frac{bl(m+2)}{(m+1)}\right] \quad [5.30]$$

$$l_1 = \frac{l_2}{2} = \frac{l}{3}$$

$$P(\sigma, V) = 1 - \exp\left[-\frac{1}{V_o}\left(\frac{\sigma_{max}}{\sigma_0}\right)^m \frac{2bdl(m+3)}{3(m+1)^2}\right] \quad [5.31]$$

$$P(\sigma, A) = 1 - \exp\left[-\frac{1}{A_o}\left(\frac{\sigma_{max}}{\sigma_o}\right)^m \frac{2bl(m+3)}{3(m+1)}\right] \quad [5.32]$$

$$l_1 = 2l_2 = \frac{2l}{3}$$

$$P(\sigma, V) = 1 - \exp\left[-\frac{1}{V_o}\left(\frac{\sigma_{max}}{\sigma_0}\right)^m \frac{2bdl(2m+3)}{3(m+1)^2}\right] \quad [5.33]$$

$$P(\sigma, A) = 1 - \exp\left[-\frac{1}{A_o}\left(\frac{\sigma_{max}}{\sigma_o}\right)^m \frac{2bl(2m+3)}{3(m+1)}\right] \quad [5.34]$$

The corresponding effective volumes and surface areas (bottom surface of the beam) are derived from equations [5.27] and [5.28] for the appropriate l_1 versus l_2 relations.

5.5. Some remarkable exact expressions for failure probability, effective volume or surface area (multiaxial elemental strength theory)

5.5.1. *Uniaxial tension*

As discussed in section 5.3, the expression of failure probability as derived from general equation [4.54] is:

$$P_v = 1 - \exp\left[-\frac{1}{V_o}.I_v(m_v, 0, 0)\left(\frac{\sigma}{\sigma_{ov}}\right)^{m_v} V\right] \quad [5.35]$$

It is worth recalling that σ_{0V} characterizes the flaw strength distribution. As such, it is different from the Weibull scale factor. Equating equation [5.35] and the corresponding Weibull equation [2.11] gives the relation between scale factors:

$$\sigma_o = \sigma_{0V}\left[I_V(m_V, 0, 0)\right]^{-\frac{1}{m_V}} \quad [5.36]$$

This equation indicates that the Weibull scale factor σ_0 is larger than the multiaxial elemental strength model one.

The effective volume as derived from general equation [5.7] is $V_E = V$.

5.5.2. *Non-uniform uniaxial stress states*

From equation [5.5], it comes for failure probability:

$$P_V = 1 - \exp\left[-\frac{1}{V_0} I_V(m_V, 0, 0) \int_V \left(\frac{\sigma_1}{\sigma_{0V}}\right)^{m_V} dV\right] \quad [5.37]$$

The expression of effective volume derived from equation [5.7] is identical to that obtained using the Weibull model:

$$V_E = \int_v \left(\frac{\sigma_1}{\sigma_{max}}\right)^{m_v} dV \quad [5.38]$$

Therefore, introducing the expressions of V_E obtained in section 5.4 for 3-point and 4-point bending into equation [5.37] gives the corresponding equations of failure probability for the multiaxial elemental strength model. It is reasonable to neglect the y–axis stress components. Results for fracture from surface are obtained in parallel.

– *3-point bending*:

$$P_v = 1 - \exp\left[-\frac{V}{V_o}\left(\frac{\sigma_{max}}{\sigma_{ov}}\right)^{m_v} \frac{1}{2(m_v+1)^2} I_v(m_v, 0, 0)\right] \quad [5.39]$$

$$P_s = 1 - \exp\left[-\frac{A}{A_o}\left(\frac{\sigma_{max}}{\sigma_{os}}\right)^{m_s} \frac{1}{m_s+1} I_s(m, 0)\right] \quad [5.40]$$

$$V_E = \frac{V}{2(m_v+1)^2} \Rightarrow K = 2(m_v+1)^2 = \frac{V_E}{V} \quad [5.41]$$

$$A_E = \frac{A}{m_s+1} \Rightarrow K = (m_s + 1) = \frac{A_E}{A} \qquad [5.42]$$

V = 4 bdl and A = 2 bl (Figure 5.1).

– *4-point bending*:

$$P_v = 1 - \exp\left[-\frac{1}{V_o}\left(\frac{\sigma_{max}}{\sigma_{ov}}\right)^{m_v}\left(\frac{2bdl_1}{m_v+1} + \frac{2bdl_2}{(m_v+1)^2}\right) I_v(m_v,0,0) \right] \qquad [5.43]$$

$$P_s = 1 - \exp\left[-\frac{1}{A_o}\left(\frac{\sigma_{max}}{\sigma_{os}}\right)^{m_s}\left(2bl_1 + \frac{2bl_2}{m_s+1}\right) I_s(m_s,0) \right] \qquad [5.44]$$

$$V_E = \frac{2bdl_1}{m_v+1} + \frac{2bdl_2}{(m_v+1)^2} \qquad [5.45]$$

$$A_E = 2bdl_1 + \frac{2bl_2}{m_s+1} \qquad [5.46]$$

The expressions for V_E and A_E obtained in section 5.3 for particular positions of forces are still valid.

5.5.3. *Multiaxial stress states: uniform stress state*

– *Equibiaxial tension*:

From equation [4.54], we obtain for principal stresses $\sigma_1 = \sigma_2$:

$$P_V = 1 - \exp\left[-\frac{V}{V_0}\left(\frac{\sigma_1}{\sigma_{0V}}\right)^{m_V} I_V(m_V,1,0) \right] \qquad [5.47]$$

From equation [5.7], we obtain for the effective volume:

$$V_E = \frac{I_v(m_v,1,0)}{I_v(m_v,0,0)} V \qquad [5.48]$$

According to values given in Table 4.1, I_v (m_v, 1, 0) ~ 0.7 and I_v (m_v, 0,0) ≈ 0.25. Consequently, $V_E \approx 2.8$ V.

It is worth pointing out that the effective volume is significantly greater than the current stressed volume. This is attributed to the effect of flaw orientation to the stress direction and of biaxial loading. The amount of flaws activated by equibiaxial stress state is larger than that under uniaxial stress state. Thus, the flaws having principal plane parallel to one stress direction that were innocuous can be critical now under stress component perpendicular to principal plane (Figure 5.3). Thus, severity is also increased for those flaws having a small angle to one stress direction (Figure 5.3).

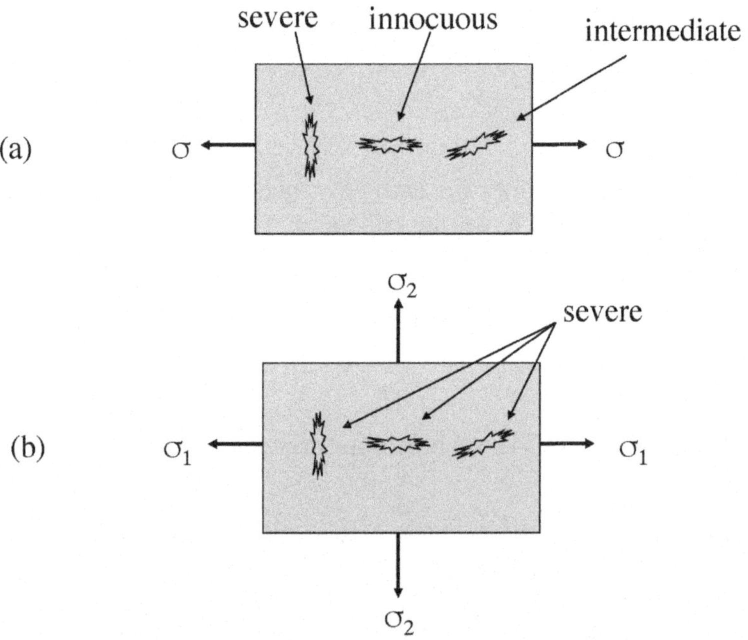

Figure 5.3. *Severity of flaws as a function of orientation to loading directions: a) uniaxial tension; b) biaxial tension*

– *Biaxial tension*:

From equation [4.54], we obtain for principal stresses $\sigma_1 < \sigma_2$:

$$P_V = 1 - \exp\left[-\frac{1}{V_0} \int_V \left(\frac{\sigma_1}{\sigma_{0V}} \right)^{m_V} I_V(m_V, \frac{\sigma_2}{\sigma_1}, 0) dV \right] \quad [5.49]$$

From equation [5.7], we obtain for the effective volume:

$$V_E = \frac{1}{I_V(m_V, 0, 0)} \left[\int_V \left(\frac{\sigma_1}{\sigma_{max}} \right)^{m_V} I_V(m_V, \frac{\sigma_2}{\sigma_1}, 0) dV \right] \quad [5.50]$$

On the basis of equations [4.48], [4.49] and [4.55], it is shown that I_v $(m_v, \frac{\sigma_2}{\sigma_1}, 0) < I_v(m_V, 1, 0)$ when $\frac{\sigma_2}{\sigma_1} < 1$. Therefore, from comparison of equations [5.48] and [5.50], it appears that V_E (biaxial) < V_E (equibiaxial), for the same volume size V.

– *Equitriaxial tension*:

From equation [4.54], we obtain for principal stresses $\sigma_1 = \sigma_2 = \sigma_3$:

$$P_V = 1 - \exp\left[-\frac{V}{V_0} \left(\frac{\sigma_1}{\sigma_{0V}} \right)^{m_V} I_V(m_V, 1, 1) \right] \quad [5.51]$$

It is demonstrated that I_v (m, 1, 1) = 1 (Figure 4.9). Thus, equation [5.51] becomes:

$$P_V = 1 - \exp\left[-\frac{1}{V_o} \left(\frac{\sigma_1}{\sigma_{0V}} \right)^{m_V} V \right] \quad [5.52]$$

As pointed out earlier in section 4.1, the scale factor in the multiaxial strength model is smaller than the Weibull's one. Equation [5.52] must not be mistaken for the Weibull one for uniaxial tension. Equation [5.52] will predict greater failure probabilities.

From equation [5.7], we obtain for the effective volume:

$$V_E = \frac{I_v(m_v,1,1)}{I_v(m_v,0,0)} V \equiv \frac{V}{I_v(m_v,0,0)} \quad [5.53]$$

According to values given in Table 4.1, I_v $(m_v,0,0)$ ~ 0.25. Consequently, $V_E \approx 4$ V.

It is worth pointing out the large value of effective volume that is quite significantly greater than the current stressed volume[1]. This large size reflects the large amount of flaws activated under triaxial stress as a result of variation in flaw orientation. Thus, the flaws having their principal plane parallel to one stress direction, which would be innocuous under uniaxial tension, are now more severe due to stress component perpendicular to principal plane (Figure 5.3). Due to equitriaxial loading, more flaws can be critical compared to equibiaxial tension. Thus, severity is also increased for more flaws having a small angle to one stress direction (Figure 5.3). The probability that a critical flaw be present increases with the number of loading directions. The effective volume is increased correlatively.

– *Triaxial tension*:

In triaxial tension, $\sigma_1 > \sigma_2 > \sigma_3 > 0$, so that:

$$\frac{\sigma_3}{\sigma_1} < \frac{\sigma_2}{\sigma_1} < 1. \quad [5.54]$$

From equation [4.54], we obtain for failure probability:

$$P_V = 1 - \exp\left[-\frac{1}{V_0}\int_V \left(\frac{\sigma_1}{\sigma_{0V}}\right)^{m_V} I_V\left(m_V, \frac{\sigma_2}{\sigma_1}, \frac{\sigma_3}{\sigma_1}\right) dV\right] \quad [5.55]$$

1 It must be recalled at this stage that the effective volume is operated on by a uniaxial stress.

From equation [5.7], we obtain for the effective volume:

$$V_E = \frac{1}{I_V(m,0,0)} \int_V \left(\frac{\sigma_1}{\sigma_{max}} \right)^{m_V} I_V \left(m_V, \frac{\sigma_2}{\sigma_1}, \frac{\sigma_3}{\sigma_1} \right) dV \qquad [5.56]$$

For a uniform triaxial stress state, expressions [5.55] and [5.56] become:

$$P_V = 1 - \exp\left[-\frac{V}{V_0} \left(\frac{\sigma_1}{\sigma_{0V}} \right)^{m_V} I_V \left(m_V, \frac{\sigma_2}{\sigma_1}, \frac{\sigma_3}{\sigma_1} \right) \right] \qquad [5.57]$$

$$V_E = \frac{I_V \left(m_V, \frac{\sigma_2}{\sigma_1}, \frac{\sigma_3}{\sigma_1} \right)}{I_V (m_V, 0, 0)} V \qquad [5.58]$$

It can be easily established that $I_v (m_v, \frac{\sigma_2}{\sigma_1}, \frac{\sigma_3}{\sigma_1}) < 1$. Thus, V_E (triaxial) < V_E (equitriaxial), for the same volume size V, which implies that the probability of existence of a critical flaw is smaller under triaxial stresses when compared to the equitriaxial case.

5.5. Conclusion

Under uniaxial stress, effective volume values range from 0 to V (tensile stressed volume). V is obtained under uniaxial tension, and 0 is obtained under compression. The size of effective volume is commensurate with severity of the stress state, which governs the probability of existence of a critical flaw. Uniaxial tension is the most severe uniaxial loading case. Compression is the less severe. The probability of fracture is 0 if the stress state is really compressive ($\sigma (x, y, z) < 0$). The stress state must not be mistaken for the remote stresses. Compressive remote stresses can induce a tensile stress state.

Under multiaxial loading, the effective volume was defined as the probabilistically equivalent size of volume under uniaxial uniform tension. Equitriaxial tension is the most severe loading case. The corresponding effective volume is the greatest. Effective volumes rank in the following order for multiaxial loading cases:

$$V_{E\ \text{equitriaxial}} > V_{E\ \text{equibiaxial}} > V$$

$$V_{E\ \text{biaxial}} < V_{E\ \text{equibiaxial}}$$

$$V_{E\ \text{triaxial}} < V_{E\ \text{equitriaxial}}$$

The ratio of the effective volume to stressed volume provides a characteristic of severity of stress state: K = $\frac{V_E}{V}$. K = 2.8 in equibiaxial loading conditions and 4 in equitriaxial loading conditions.

6

Size and Stress-state Effects on Fracture Strength

6.1. Introduction

The term "size effects" designates the dependence of fracture resistance on the size of stressed volume or surface. The larger the size, the smaller the fracture resistance, since the probability that a severe flaw is present increases with size. As flaw severity depends on the stress state, the stress state contributes to size effects. The effect of stress state on fracture can be characterized also using the effective volume or surface concept. As discussed in Chapter 5, the bigger the effective volume or surface, the most severe the stress state, and the smaller the fracture strength. The statistical-probabilistic approaches to fracture enable us to quantify and predict the size effects on fracture strengths. In this chapter, approaches based on the Weibull and the multiaxial elemental strength models are detailed.

6.2. Uniform uniaxial stress state

6.2.1. *Effects of stressed volume or surface size on strengths*

Let us consider two samples of a material, having volumes $V_1 > V_2$, subjected to unixial tension. The probability of fracture is given by (Weibull model):

$$P_1(\sigma_1, V_1) = P_1 = 1 - \exp\left[-\frac{V_1}{V_0}\left(\frac{\sigma_1}{\sigma_0}\right)^m\right] \qquad [6.1]$$

$$P_2(\sigma_2, V_2) = P_2 = 1 - \exp\left[-\frac{V_2}{V_0}\left(\frac{\sigma_2}{\sigma_0}\right)^m\right] \qquad [6.2]$$

where σ_1 and σ_2 are the stresses operating on volumes V_1 and V_2, respectively. From equations [6.1] and [6.2], the following relation between σ_1 and σ_2 is derived:

$$\frac{\sigma_2}{\sigma_1} = \left(\frac{V_2}{V_1}\right)^{-1/m}\left(\frac{Ln(1-P_1)}{Ln(1-P_2)}\right)^{1/m} \qquad [6.3]$$

For an appropriate comparison of stresses, the same values of failure probability $P_1 = P_2$ are considered. The right side of equation [6.3] allows determination of the value of stress σ_2 that corresponds to σ_1 when the volume changes from V_1 to V_2:

$$\frac{\sigma_2}{\sigma_1} = \left(\frac{V_2}{V_1}\right)^{-1/m} \qquad [6.4]$$

According to equation [6.4], the strength is commensurate with the reciprocal of volume. Size effects depend on value of Weibull modulus. The smaller the Weibull modulus, the larger the strength dependence on size.

The following equations of failure probability are derived from equation [4.54] of the multiaxial elemental model for a uniform uniaxial stress state:

$$P_1 = 1 - \exp\left[-\left(\frac{\sigma_1}{\sigma_0}\right)^m I_v(m,0,0)\frac{V_1}{V_0}\right] \qquad [6.5]$$

$$P_2 = 1 - \exp\left[-\left(\frac{\sigma_2}{\sigma_0}\right)^m I_v(m,0,0)\frac{V_2}{V_0}\right] \qquad [6.6]$$

with $I_v(m, 0, 0) = \int_0^{\pi/2} \cos^m\theta \sin\theta(1+4\sin^2\beta\cos^2\theta)^{m/4} d\theta$ (equation [4.68]).

From equations [6.5] and [6.6], the same size effect relation as above is obtained ($P_1 = P_2$):

$$\frac{\sigma_2}{\sigma_1} = (\frac{V_2}{V_1})^{-1/m} \qquad [6.7]$$

When fracture occurs from the outer surface of solid, equations similar to equations [6.4] and [6.7] are obtained whatever the model (Weibull or multiaxial elemental strength model):

$$\frac{\sigma_2^s}{\sigma_1^s} = \left(\frac{A_2}{A_1}\right)^{-\frac{1}{m_s}} \qquad [6.8]$$

where A_1 and A_2 are the corresponding areas of outer surfaces, m_s is the shape parameter of specimen (Weibull) or flaw (multiaxial elemental strength model) strength distributions.

Figure 6.1 illustrates the effect of volume size on fracture resistance. It is shown that size effects on strengths increase with decreasing shape parameter value (m), i.e. when the flaw population is highly heterogeneous in terms of severity. By contrast, homogeneous materials which are characterized by large m values are quite insensitive to size effects.

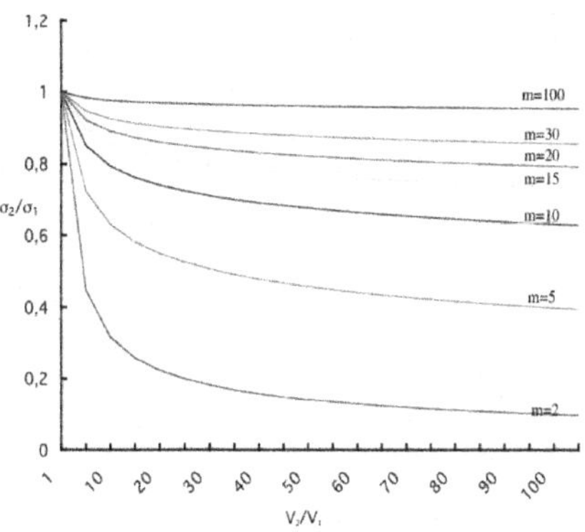

Figure 6.1. *Size effect on fracture strength under uniform uniaxial tension for various values of Weibull modulus. For a color version of the figure, see www.iste.co.uk/lamon/brittle.zip*

6.2.2. *Respective effects of volume and surface on fracture*

The probabilities of fracture from flaws located at the surface and from those located within the interior of a specimen are given by the following Weibull equations:

$$P_V = 1 - \exp\left[-\frac{V}{V_0}\left(\frac{\sigma_v}{\sigma_{ov}}\right)^{m_v}\right] \qquad [6.9]$$

$$P_S = 1 - \exp\left[-\frac{A}{A_o}\left(\frac{\sigma_s}{\sigma_{os}}\right)^{m_s}\right] \qquad [6.10]$$

where the subscripts v and s refer to surface and volume.

In the presence of concurrent flaw populations, failure probability is derived from the product of survival probabilities according to the weakest link principle (Chapter 2 section 2.2.1):

$$P = 1 - (1 - P_s)(1 - P_v) \qquad [6.11]$$

The condition $P_S = P_V$ marks the transition between fracture from one flaw population to fracture from the other one. The transition strength σ^* is derived by equating equations [6.9] and [6.10] for $\sigma_s = \sigma_V$:

$$\sigma^* = \left(\frac{V}{A}\frac{A_0}{V_0}\frac{\sigma_{0S}{}^{m_s}}{\sigma_{0V}{}^{m_v}}\right)^{\frac{1}{m_s - m_v}} \qquad [6.12]$$

Note that equation [6.12] is not valid when $m_V = m_S$.

The preponderant flaw population is characterized by the smaller strengths and associated fracture probabilities in the cumulative distribution. The domination of one population is indicated by the value taken by the following ratio derived from equations [6.9] and [6.10]:

$$y = \frac{Ln(1 - P_S)}{Ln(1 - P_v)} = \frac{A}{V} \cdot \frac{V_o}{A_o}\left(\frac{\sigma_s}{\sigma_{os}}\right)^{m_s}\left(\frac{\sigma_{ov}}{\sigma_v}\right)^{m_v} \qquad [6.13]$$

When $y > 1$, $P_s < P_v$ in the cumulative distribution of flaw strengths, which indicates that fracture is dominated by the population of surface-located flaws. Conversely, when $y < 1$ ($P_s > P_v$), fracture starts from the volume-located flaws.

In the multiaxial elemental strength model, probabilities of fracture from, respectively, volume-located and surface-located flaws are given by the following equations:

$$P_v = 1 - \exp[-\frac{V}{V_0} I_v(m_v, 0, 0)(\frac{\sigma_v}{\sigma_{ov}})^{m_v}] \quad [6.14]$$

$$P_s = 1 - \exp[-\frac{A}{A_o} I_s(m_s, 0, 0)\left(\frac{\sigma_s}{\sigma_{os}}\right)^{m_s}] \quad [6.15]$$

with $I_s = \frac{2}{\pi}\int_0^{\frac{\pi}{2}} \cos^m \psi(1+4\cos^2\psi\sin^2\psi)^{m/4} d\psi$.

I_v is given by equation [4.68].

The expression for y which identifies the preponderant population is now:

$$y = \frac{A}{V} \cdot \frac{V_o}{A_o}\left(\frac{\sigma_s}{\sigma_{os}}\right)^{m_s}\left(\frac{\sigma_{ov}}{\sigma_v}\right)^{m_v} \frac{I_s(m_s, 0)}{I_v(m_v, 0, 0)} \quad [6.16]$$

According to Table 4.1, $I_s(m_s,0)$ is generally larger than $I_v(m,0,0)$. The values of $I_s(m_s,0)/I_v(m_v,0,0)$ range between 1.25 and 2.13 depending on m value. Thus, the comparison with equation [6.14] indicates that the contribution of surface-located flaws is more significantly accounted for by the multiaxial elemental strength model.

Similarly, the transition stress differs from the Weibull model one by the ratio $\frac{I_s(m_s, 0)}{I_v(m_v, 0, 0)}$:

$$\sigma^* = [\frac{V}{A} \cdot \frac{A_o}{V_o} \cdot \frac{I_s(m_s,0)}{I_v(m_v,0,0)} \cdot \frac{\sigma_{ov}^{m_v}}{\sigma_{os}^{m_s}}]^{1/m_v - m_s} \qquad [6.17]$$

Size effects on the transition strength can be derived from the ratio of transition strengths for different volume sizes and surface areas. From equation [6.12] or [6.17], we obtain:

$$\frac{\sigma_2^*}{\sigma_1^*} = \left[\frac{V_2}{V_1}\frac{A_1}{A_2}\right]^{\frac{1}{m_s - m_v}} \qquad [6.18]$$

This relation is obtained with both the Weibull and the multiaxial elemental strength models.

6.3. Non-uniform uniaxial stress state

Using effective volume (V_E) or surface area (A_E) is a convenient way to establish size effect laws in the case of non-uniform stress state. The corresponding Weibull equations are:

$$P_1 = P(\sigma_1, V_1) = 1 - \exp [\frac{V_{E1}}{V_0} \cdot (\frac{\sigma_1}{\sigma_0})^m] \qquad [6.19]$$

$$P_2 = P(\sigma_2, V_2) = 1 - \exp [\frac{V_{E2}}{V_0} \cdot (\frac{\sigma_2}{\sigma_0})^m] \qquad [6.20]$$

where σ_2 and σ_1 are the maximum stresses operating on volumes V_2 and V_1, respectively.

Equating equations [6.20] and [6.21] yields the following relation for size effects on strengths:

$$\frac{\sigma_2}{\sigma_1} = \left[\frac{V_{E2}}{V_{E1}}\right]^{-\frac{1}{m}} \qquad [6.21]$$

Note that equation [6.21] reduces to equation [6.8] for uniform tensile stress state ($V_E = V$). Size effects depend on value of Weibull modulus. The strength is commensurate with the reciprocal of effective volume size modulated by the Weibull modulus value:

$$\sigma.V_E^{\frac{1}{m}} = k \qquad [6.22]$$

where k is a constant.

As shown in Chapter 5, the effective volume size is related to stressed volume size by the severity factor K:

$$V_E = K.V \qquad [6.23]$$

K depends on stress state. K tends to 1 as the stress state becomes uniform uniaxial. Hereafter are expressions obtained for a few simple stress states (Chapter 5):

$K_t = 1$ for uniform tension

$$K_{3p} = \frac{1}{2(m+1)^2} \text{ for 3-point bending} \qquad [6.24]$$

$K_{4p} = \frac{m+3}{6(m+1)^2}$ for 4-point bending ($2\,l_1 = l_2$)

$K = \frac{1}{2(m+1)}$ for pure bending

Thus, the effective volume size is governed by the size of tensile stressed volume and stress state. K is a function of Weibull modulus. K generally decreases with increasing Weibull modulus as shown in Figure 6.2.

For a given loading mode, the ratio of effective volumes is identical to the ratio of stressed volume sizes:

$$\frac{V_{E2}}{V_{E1}} = \frac{V_2}{V_1} \qquad [6.25]$$

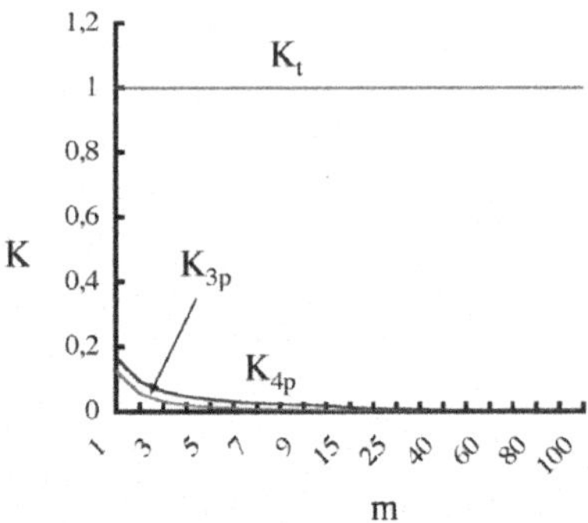

Figure 6.2. *Effect of loading mode on fracture strength: values of severity factor K = V_E/V as a function of Weibull modulus m for uniaxial uniform tension (K_t), 3-point bending (K_{3p}) and 4-point bending (K_{4p}). For a color version of the figure, see www.iste.co.uk/lamon/brittle.zip*

Size effects on strengths are governed by the ratio of stressed volume sizes (equation [6.8]).

When stressed volume size is constant, strength depends on loading mode. From equations [6.21] and [6.23], we obtain:

$$\frac{\sigma_2}{\sigma_1} = (\frac{K_2}{K_1})^{-1/m} \qquad [6.26]$$

According to equation [6.26], the strength is commensurate with the reciprocal of K. The smaller the K, the steeper is the stress gradient. Like size effect, the stress state effect is favored by small m values. In other words, those materials with heterogeneous flaw populations have stress state sensitive strength.

Failure probability of the *multiaxial elemental strength model* is given by the following equations for a non–uniform stress state (Chapter 5, equation [5.37]):

$$P_1 = 1 - \exp [-\frac{1}{V_0}\int_v (\frac{\sigma}{\sigma_0})^m I_v(m,0,0)dV] \qquad [6.27]$$

$$= 1 - \exp [-\frac{V_E}{V_0}\left(\frac{\sigma_{max}}{\sigma_0}\right)^m I_v(m,0,0)] \qquad [6.28]$$

where V_E is the effective volume and σ_{max} is the peak stress. The effective volume is given by equation [5.38] (Chapter 5). As above, equating failure probabilities for different stressed volume sizes, it comes from equation [6.28]:

$$\frac{\sigma(V_2)}{\sigma(V_1)} = \left(\frac{V_{E2}}{V_{E1}}\right)^{-1/m} \qquad [6.29]$$

$$\frac{\sigma(V_2)}{\sigma(V_1)} = \left(\frac{K_2 V_2}{K_1 V_1}\right)^{-1/m} \qquad [6.30]$$

6.4. Multiaxial stress state: multiaxial elemental strength model

According to equation [5.6], failure probability for effective volumes V_{E1} and V_{E2} is given by equations:

$$P(V_1) = 1 - \exp\left[-\frac{1}{V_o} I_v(m_v, 0, 0)\left(\frac{\sigma^1_{max}}{\sigma_{ov}}\right)^m V_E^1\right] \qquad [6.31]$$

$$P(V_2) = 1 - \exp\left[-\frac{1}{V_o} I_v(m_v, 0, 0)\left(\frac{\sigma^2_{max}}{\sigma_{ov}}\right)^m V_E^2\right] \qquad [6.32]$$

Equating failure probabilities gives the strength–effective volume relation:

$$\frac{\sigma^2_{max}}{\sigma^1_{max}} = \left(\frac{V_{E2}}{V_{E1}}\right)^{-1/m} \qquad [6.33]$$

Equation [6.33] indicates the same trend as above: the strength is commensurate with the reciprocal of effective volume size modulated by the shape factor m. The equation of effective volume is generally complex (equation [5.7]). As discussed in Chapter 5, it must be calculated numerically, except for those simple cases when analytical expressions of stress state can be developed.

For uniform multiaxial stress state, the effective volume is given by the following expression, according to equation [5.9]:

$$\mathrm{V_E} = \frac{I_v(m_v, \frac{\sigma_2}{\sigma_1}, \frac{\sigma_3}{\sigma_1})}{I_v(m_v, 0, 0)} \mathrm{V} = \mathrm{KV} \qquad [6.34]$$

Consequently, the same laws as above govern sizes effects on strength:

$$\frac{\sigma_{max}^2}{\sigma_{max}^1} = \left(\frac{K_2 V_2}{K_1 V_1}\right)^{-\frac{1}{m}} \qquad [6.35]$$

For given loading mode ($K_2 = K_1$), size effects on strength depend only on the size of stressed volume:

$$\frac{\sigma_{max}^2}{\sigma_{max}^1} = \left(\frac{V_2}{V_1}\right)^{-\frac{1}{m}} \qquad [6.36]$$

When stressed volume is constant, strength depends on severity factor K:

$$\frac{\sigma_{max}^2}{\sigma_{max}^1} = \left(\frac{K_2}{K_1}\right)^{-\frac{1}{m}} \qquad [6.37]$$

6.5. Applications

6.5.1. *Influence of loading conditions*

The dependence of fracture strength on loading conditions is exemplified on tests that are commonly used for the determination of mechanical strength of materials, i.e. traction, 3-point and 4-point bending that generate essentially uniaxial stress states. For identical stressed volume size, the 3-point (σ_{3p}), 4-point (σ_{4p}) bending and tensile strengths (σ_t) can be compared by the following ratios derived from expressions for K given by equations [6.24]:

$$\frac{\sigma_{3p}}{\sigma_t} = [2(m+1)^2]^{1/m} \quad [6.38]$$

$$\frac{\sigma_{4p}}{\sigma_t} = [\frac{6(m+1)^2}{m+3}]^{1/m} \quad [6.39]$$

$$\frac{\sigma_{3p}}{\sigma_{4p}} = [\frac{1}{3}(m+3)]^{1/m} \quad [6.40]$$

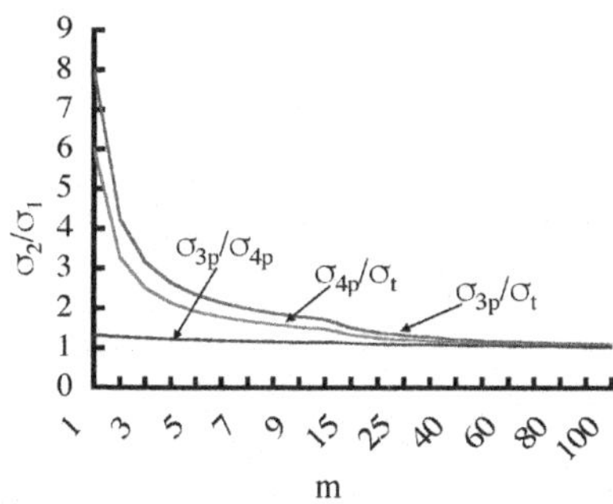

Figure 6.3. *Effect of uniaxial loading modes on the fracture strength for constant volume: 3-point bending (σ_{3p}), 4-point bending (σ_{4p}) and uniform tension (σ_t). For a color version of the figure, see www.iste.co.uk/lamon/brittle.zip*

Figure 6.3 shows that traction is much more severe than bending for m values smaller than about 15. The ratios of bending to tensile strengths increase when m approaches values as small as 1, i.e. when material possesses flaws with significant scatter in severity. With homogeneous materials (m takes very large values (m > 1000)), the ratios of bending to tensile strengths tend to 1. 3-point bending is less severe than 4-point bending as a result of the presence of uniform stresses in the inner span volume of 4-point bending specimens. Traction is the most severe test for heterogeneous materials because all the flaws are subjected to maximum stress. It can be deduced from the trends shown by this example that the strength will depend on the size of volumes under uniform stresses versus those subjected to stress gradients. Traction and 3-point bending are the extreme cases for uniaxial stress states. In the presence of a given population of flaws, a vertical stress gradient will be the less severe loading case. This case should not be mistaken for the stress concentration caused by a sharp notch that would not pertain to the population of pre-existing flaws in the material. The notch would induce much larger stresses than the applied load.

It is interesting to compare the influence of *multiaxial stress states* to that of uniaxial stress states, which may correspond to the strength of a component versus that of a test specimen. These strengths are never compared, because carrying multiaxial tests is unusual for practical difficulties, and because sound models for prediction of fracture under multiaxial stress states are required. For simplicity, the discussion is limited to the simple typical cases when closed-form equations of stress states are available and effective volume can be easily calculated. These loading conditions are the most severe and the effective volume is maximal, as shown in Chapter 5. For identical stressed volume sizes, the equitriaxial and equibiaxial tensile strengths (σ_{eqt} and σ_{eqb}) are related to uniaxial tensile (σ_t) and 3-point bending (σ_{3p}) strength by the following formulae derived from effective volume expressions established in Chapter 5:

$$\sigma_{eqb} / \sigma_t = \left[\frac{I_v(m,1,0)}{I_v(m,0,0)} \right]^{-\frac{1}{m}} \quad [6.41]$$

$$\sigma_{eqb} / \sigma_{3p} = \left[\frac{I_v(m,1,0)}{I_v(m,0,0)} 2(m+1)^2 \right]^{-\frac{1}{m}} \quad [6.42]$$

$$\sigma_{eqt} / \sigma_t = \left[\frac{1}{I_v(m,0,0)} \right]^{-\frac{1}{m}} \quad [6.43]$$

$$\sigma_{eqt}/\sigma_{3p} = \left[\frac{2(m+1)^2}{I_v(m_v,0,0)}\right]^{-\frac{1}{m}} \quad [6.44]$$

Figure 6.4 shows that multiaxial strengths are smaller than uniaxial strengths. This trend reflects the influence of flaw orientation to stress direction as discussed in Chapters 2 and 3. The ratio of multiaxial/uniaxial strengths increases with the value of shape parameter m. Thus, for m = 5, the values of multiaxial strength σ_{eqb} and σ_{eqt} are 30% of the uniaxial flexural strength σ_{3p}. Then, for m = 20, they are 60% of σ_{3p}. These examples clearly demonstrate the importance of material weakening induced by multiaxial loading modes, which implies that relationships between component strength in service and material strength measured using laboratory tests on specimens must be considered with much attention.

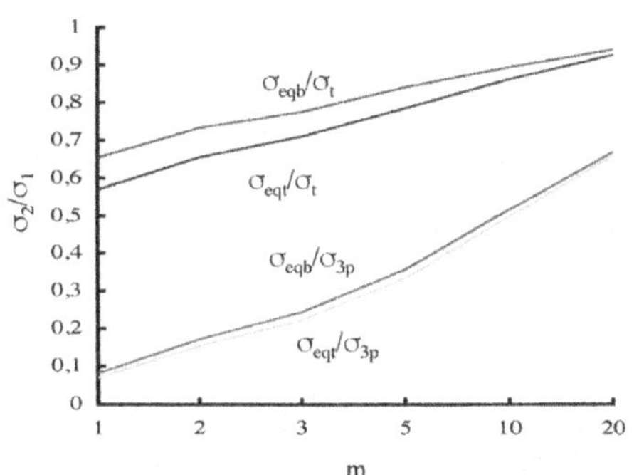

Figure 6.4. *Effect of multiaxial loading modes on the fracture strength for constant volume: equibiaxial tension (σ_{eqb}) and equitriaxial tension (σ_{eqt}) versus uniaxial tension (σ_t) and 3-point bending (σ_{3p}). For a color version of the figure, see www.iste.co.uk/lamon/brittle.zip*

6.5.2. *Importance of stress effects on the fracture of fiber reinforced ceramic matrix composites*

Effects of stress state govern the ultimate fracture of these ceramic composites through the strength of filaments carrying the applied loads. Thus, for those composites with constituents having contrasting Young's moduli such as $E_m \geq E_f$, the matrix carries a part of the load, so that the fibers are underloaded. In the presence of stress-induced damage by cracks in the matrix and the fiber/matrix interfaces, the load is carried locally by the fibers alone, which induces a stress gradient. The fibers are thus subjected to a non-uniform tensile stress state made up of alternate domains of uniform and non-uniform stresses (Figure 6.5):

– those domains where the applied load is carried by the fibers alone, as a result of matrix crack opening displacement. The stress on fibers is maximum;

– those domains located in the vicinity of debond, where the stress on fiber decreases;

– those domains where the interface is intact. Load sharing is effective, so that the stresses on fibers are uniform and minimum.

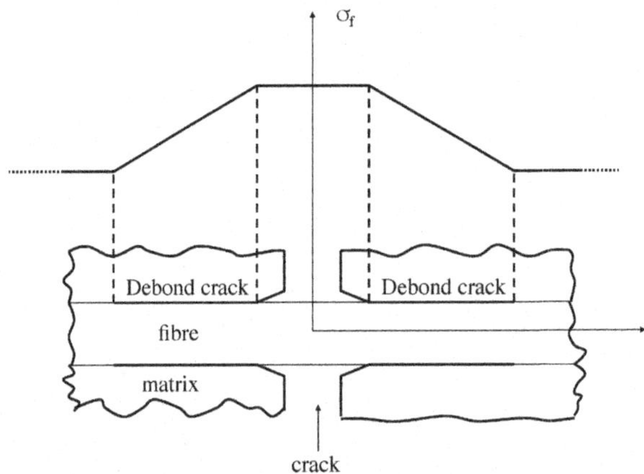

Figure 6.5. *Stress (σ_f) operating on fiber in the vicinity of a matrix crack as a result of shear lag induced by fiber/matrix debonding, in a ceramic matrix composite subject to uniaxial tension parallel to fiber axis*

According to section 6.5.1, the severity factor $K = V_E/V < 1$ due to the presence of stress gradients. Consequently, the actual fiber strength is greater than the fracture stress in uniform tension. Generally, under load increasing monotonously, matrix cracking and associated fiber debonding progress, so that the stress-on-fiber tends to a uniform state that is reached when the debond cracks have coalesced. Thus, K increases to 1 so that actual fiber strength decreases to minimum given by the tensile strength under uniform stress. Thus, the fiber within the composite does not fracture while K < 1. Uniform tension is the ultimate stress state on fiber that is required for fracture of those composites such that $E_m > E_f$. Unless higher stresses operate locally for some reason (artifact such as too strong fiber/matrix bond, specimen overloading), ultimate fracture cannot occur while fibers are still bonded to the matrix.

The severity factor $K_f = \frac{V_E}{V_F}$ (V_F is the volume of fiber) for a microcomposite reinforced by a single fiber and subjected to uniaxial stress is given by the following equation established using the method discussed in Chapter 5. Details are given in Chapter 10 and in [GUI 96, LIS 97b, LAM 09].

$$K_f(n) = \frac{1 + 2n\frac{ld}{L}A_f}{(1+a)^m} \qquad [6.45]$$

where n is the number of matrix cracks perpendicular to fiber axis; l_d is the length of the interface crack resulting from deflection of matrix crack; L is microcomposite specimen length; a is the load sharing factor, $a = \frac{E_m V_m}{E_f V_f}$ with E_m and E_f the Young's moduli of the matrix and the fiber, respectively, V_m and V_f are the volume fractions of matrix and fiber, respectively; m is the fiber Weibull modulus.

The expression of A_f is:

$$A_f \approx \beta(1+a)^m - \frac{1-\beta}{m+1}\frac{1-(1+a)^{m+1}}{a} - 1 \qquad [6.46]$$

$\beta = \frac{l_0}{l_d}$. l_0 is the debond length without fiber/matrix contact (for instance, as a result of Poisson fiber lateral contraction). Figure 6.5 shows the stress state operating on fiber in the vicinity of matrix and interface cracks. Such microcomposite configuration is the unit cell of composite material. The Weibull analysis was used for simplicity.

In the absence of crack in the matrix, n = 0 and

$$K_f(0) = \frac{1}{(1+a)^m} \qquad [6.47]$$

For an SiC/SiC composite (made of SiC matrix reinforced by SiC fibers), $E_f \approx 200$ GPa, $E_m \approx 400$ GPa, $V_f \approx 0.4$ and $V_m \approx 0.6$, so that $a \approx 2$ and $K_f(0) \approx 4.10^{-3}$ for m = 5. Consequently, according to above, the small value of $K_f(0)$ indicates that the actual strength of fiber in the uncracked microcomposite specimen is much larger than the tensile strength of fiber under uniform tension. According to equation [6.30], the initial strength ratio is given by:

$$\frac{\sigma_2}{\sigma_1} = \left[K_f(0)\right]^{-1/m} = 1 + a \qquad [6.48]$$

where σ_2 is the actual fiber strength in the microcomposite specimen, σ_1 is the strength of fiber under uniform tension. In the typical case of SiC/SiC microcomposite, the initial value of strength ratio is $\sigma_2/\sigma_1 = 3$.

After saturation of matrix cracking, the debond cracks tend to merge. At this stage, $n = n_s$ and $2n_s l_d = L$. The corresponding severity factor $K_f(n_s)$ is now:

$$K_f(n_s) = \frac{1+A_f}{(1+a)^m} = \beta + \left(\frac{\beta-1}{m+1}\right)\left(\frac{1+a}{a}\right) \qquad [6.49]$$

$K_f(n_s)$ is an increasing function of β: $K_f(n_s) < 1$ while $\beta < 1$, $K_f(n_s) = 1$ when $\beta = 1$, which shows that the severity factor is less than maximum value, and indicates that the applied stress does not reach the tensile strength as far as the fiber is not totally debonded. It is thus shown that, as a result of stress state effect on strength, fiber fracture is preceded by matrix damage, which permits the stress-on-fiber to exceed fiber strength.

The opposite phenomenon is observed when matrix Young's modulus is much smaller than that of fiber ($E_m << E_f$). The corresponding load sharing factor a = 0, and the severity factor $K_f(0) = 1$ (equation [6.47]). As a result, the actual fiber strength coincides with the tensile strength under uniform tension from the beginning of loading. Matrix damage is not a prerequisite to fiber fracture.

This phenomenon is evidenced on polymer matrix reinforced by carbon fiber having the following Young's moduli: $E_m \approx 5$ GPa, $E_f \approx 200$ GPa. The corresponding value is $a \approx 25.10^{-3}$ for $V_f = V_m = 0.5$. According to equation [6.47], $K_f(0) = 1$, which indicates that fracture can occur independently of matrix damage It is well known that fracture of fibers occurs first. Then, K_f is unchanged in the presence of broken fibers. Furthermore, fiber fractures can occur as far as integrity of the composite is kept. This phenomenon is expected in carbon/carbon composites reinforced with crutinuous high modulus fibers.

6.5.3. *Influence of volume or surface size: disadvantages and benefits*

Figures 6.1–6.4 and Figure 6.6 illustrate the significance of strength decay induced by changes in specimen size or loading mode. This effect is marked for small m values (heterogeneous materials) and mild when m approaches large values (deterministic material). The strength of a component tends to be smaller than that of a test specimen, depending on respective stress states and dimensions. Consequently, size and stress state effects should be accounted for in component design computations based on laboratory material data. Furthermore, they should also be accounted for in material data sheets. The data are useless when specimen dimensions and testing conditions are not given.

The high resistance to fracture of small bodies (such as brittle fibers and ceramic layers) and materials made up of small constituents (such as multilayers, fiber reinforced ceramic matrix composites, etc.) can also be attributed to size effects on constituent strengths. Figure 6.6 shows the strength increases for a brittle material (calculated using equation [6.4]) when the volume decreases from the meter to the nanometer length scale. The corresponding gain is outstanding. For m ~ 10, as commonly observed on ceramics, the strength becomes 6 times larger when the volume decreases from centimeter to micrometer length scale[1], and 25 times larger when volume decreases from centimeter to nanometer length scale. Advanced materials benefit from scale effect, when the length scale of constituents tends to micrometer or nanometer. The term "constituents" designates here elements having a reinforcing function, such as fibers in composites or layers in multilayers. Thus, the strength of a ceramic fiber with diameter of a few micrometers is 6 times greater than that of corresponding bulk material with a few centimeters size.

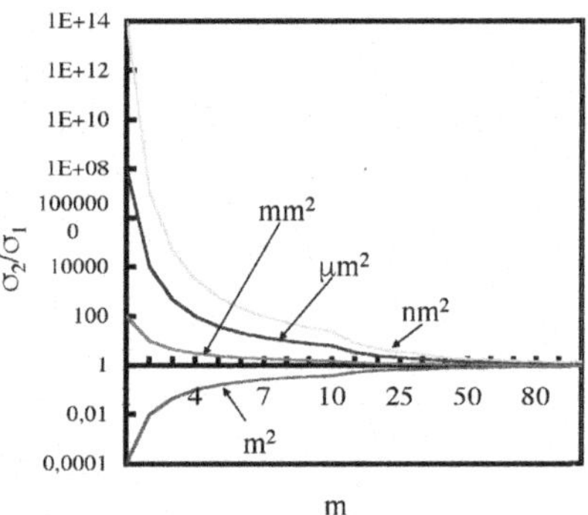

Figure 6.6. *Size effect on strength: fracture strength in uniaxial tension at various length scales (volumes V_2) with respect to a reference volume V_1 having cross–sectional area of 1 cm^2. Volumes V_2 have cross-sectional areas of 1m^2, 1 mm^2, 1 μm^2 or 1 nm^2, and the same length as V_1. For a color version of the figure, see www.iste.co.uk/lamon/ brittle.zip*

6.5.4. *Influence of shape and geometry: effects of surface-located and volume-located flaw populations*

In this section, trends in the influence of solid shape and geometry on size effects are highlighted on the particular case of uniform stress state. The case of non-uniform or multiaxial stress state can be treated by replacing V and A by equivalent volume and surface area. Then,

1 The length is taken to be constant, whereas the cross-sectional area is changed.

exact values of V_E and A_E are required for the analysis of specific loading cases (Chapter 5).

The influence of shape and geometry is governed by the competition between surface- and volume-located flaw populations. The case of concurrent populations is considered here. The volume and the outer surface area of solids are not independent, they are proportional so that the relation between both can be written as: $\lambda = \frac{V}{A}$. For instance, for a cylinder with radius r, $\lambda = r/2$ (when considering the outer surface $A = 2\pi rl$, which contributes to fracture under tensile loading in the direction of cylinder axis). For a beam having a square section $\lambda = \alpha/4$ (with $A = 4\alpha l$) a with d the edge length.

Introducing λ in equation [6.18], it comes for the ratio of transition strengths:

$$\frac{\sigma_2^*}{\sigma_1^*} = \left[\frac{\lambda_2}{\lambda_1}\right]^{\frac{1}{m_s - m_v}} \qquad [6.50]$$

with $m_s \neq m_v$. The following trends can be deduced from equation [6.50]:

– when λ remains constant, the transition strength is unchanged. As the volume and the surface increase, they both tend to weaken. As the corresponding strengths decrease, the contribution of preponderant population is enhanced, according to Figure 6.7. The preponderant population consists of either surface-located or volume-located flaws. As shown in Figure 6.7, the strength distributions are shifted toward low strengths;

– when $m_s > m_v$, σ* increases with λ. This means that the contribution of preponderant population is enhanced by shape changes (Figure 6.8(a));

– when $m_v > m_s$, σ* decreases with increasing λ. The contribution of stronger population (at high strengths in the statistical distribution) will be enhanced by shape changes that cause λ increase Figure 6.8(b)).

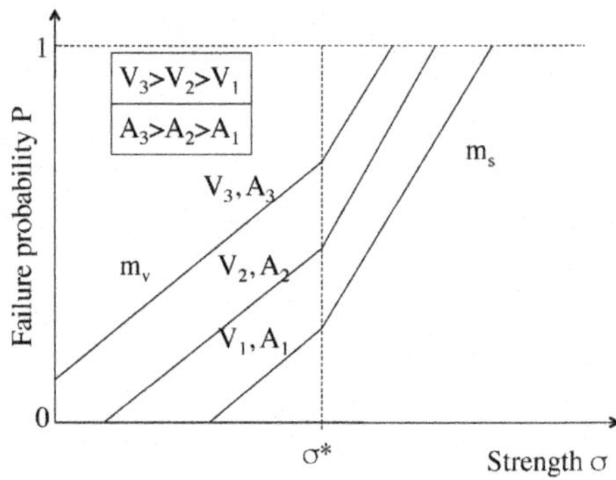

Figure 6.7. *Size effect on strengths in the presence of concurrent populations of surface-located and volume-located flaws: schematic diagram showing the influence of volume and surface area changes on strength distribution (linearized plot) when the transition strength remains constant (λ = constant). The volume-located flaw population is considered to be the preponderant one*

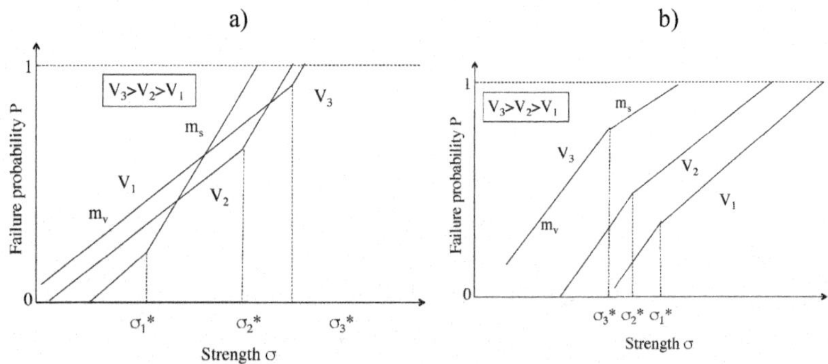

Figure 6.8. *Size effect on strengths in the presence of concurrent populations of surface-located and volume-located flaws: schematic diagram showing the influence of volume and surface area changes on strength distribution (linearized plot) when the transition strength a) increases with λ ($m_S>m_V$), and b) decreases with increasing λ ($m_S<m_V$). The volume-located flaw population is assumed to be the preponderant one*

Figure 6.9 illustrates the influence of respective values of m_s and m_v on transition strength.

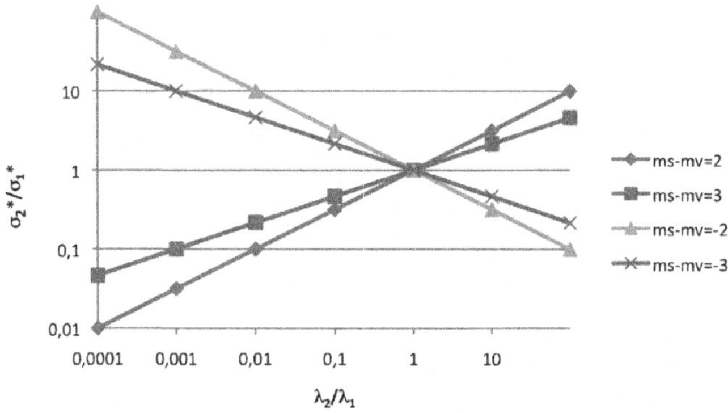

Figure 6.9. *Size effect on transition strength in the presence of concurrent populations of surface-located and volume-located flaws: influence of λ and values of m_S–m_V*

The presence of concurrent surface-located and volume-located flaws has important implications for failure prediction using data measured on laboratory specimens. The contribution of the population that is preponderant on laboratory tests may become negligible or be enhanced depending on the values of λ and shape parameters. As a consequence, it is clear that the strength data obtained on test specimens may not correspond to the population that does govern the fracture of component. Failure cannot be properly predicted if data on both populations are not documented nor used, which implies that laboratory tests need to be selected accordingly.

Figure 6.10 exemplifies the trend of Figure 6.8 on actual silicon nitride specimens, for the following experimental values of statistical parameters reported in Chapter 7:

$m_s = 8$, $\sigma_{0s} = 300$ MPa

$m_v = 11$, $\sigma_{0v} = 235$ MPa

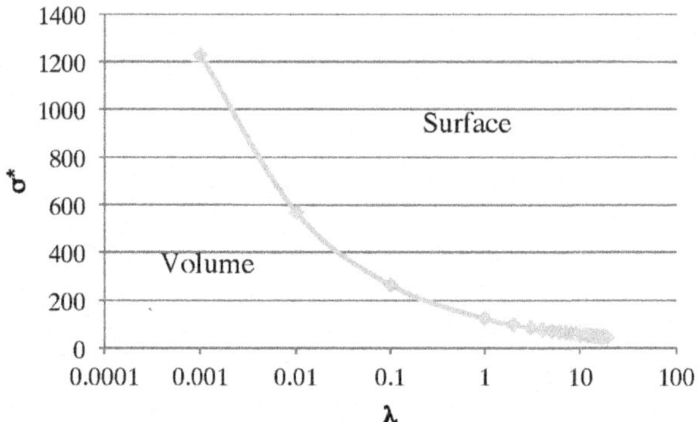

Figure 6.10. *Size effect on transition strength in the presence of concurrent populations of surface-located and volume-located flaws: influence of λ in the case of actual silicon nitride ($m_S - m_V = -3$)*

It appears that σ^* decreases with increasing λ which means that the contribution of the preponderant population (volume-located flaws) decreases, whereas the contribution of the surface population increases.

This trend illustrates perfectly the issue of specimen-component strengths relationships. The strength on small Si_3N_4 samples in uniform tension will be governed by volume-located flaws, whereas the strength of larger components will be dictated by surface-located flaws. As a result, important errors can be introduced in component failure predictions if this trend is not taken into account and if appropriate strength data are not available and introduced in the equations.

Opposite trend in failure would be obtained with $m_v > m_s$ which shows the significant influence of scatter in flaw strengths (Figure 6.11).

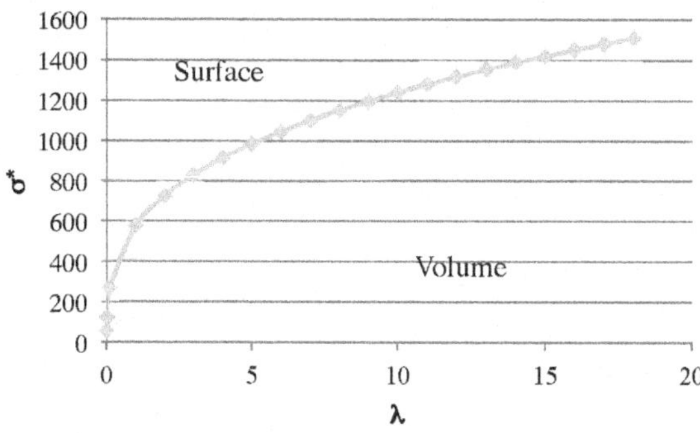

Figure 6.11. *Size effect on transition strength in the presence of concurrent populations of surface-located and volume-located flaws: influence of λ in the particular case of silicon nitride if the value of $m_S - m_V = 3$*

6.6. Conclusion

The statistical-probabilistic models provide a comprehensive insight into the dependence of fracture strength on specimen size and stress state. The governing laws obtained using the Weibull and the multiaxial elemental strength models show that the fracture strength is dictated by the reciprocal of effective volume or outer surface area and by shape parameter, which is either the Weibull modulus or the shape parameter of the statistical distribution of flaw strengths. Similar equations are obtained with both models for uniaxial loading. However, the equations of effective volume or surface area involve the I-functions that account for flaw orientation versus stresses in the multixial elemental strength model (Chapter 4).

Uniform tension is the most severe among the uniaxial loading modes. However, equibiaxial and equitriaxial stress states are comparatively more severe, as the flaws are loaded equally in several directions. Heterogeneous materials that contain broad populations of flaws with wide variability in strengths are the most sensitive to size and stress state effects. Sensitivity decreases with decreasing variability in flaw strength.

Size and stress state effects on strength play a significant role in various issues such as the damage and fracture of fiber reinforced ceramics, the performances of brittle materials having a structure combining brittle elements (multilayers, etc.), test specimen-component strength relationship, volume versus surface and shape versus geometry influences. The trends become more complex in the presence of concurrent surface-located and volume-located flaw populations. Their respective contributions to failure depend on ratio of volume and surface area and on respective values of shape parameters. The contribution of that population that was preponderant on laboratory tests on small size specimens may be either enhanced or reduced as volume and surface area increase. Specific cases encountered in component design can be investigated using the general equations proposed in this chapter.

7

Determination of Statistical Parameters

7.1. Introduction

The statistical parameters of interest in statistical-probabilistic approaches to fracture index the distribution of strength data, i.e. the flaw strengths for the physics-based models, or the strengths of the specimens for the Weibull model. These parameters can be regarded as characteristics of populations of flaws or specimens, and consequently, of materials. They include a dispersion parameter or scale parameter, and a shape parameter. The shape parameter is the exponent of failure characteristic in the power law-based approaches. Computation of failure probability of a structural component made of a brittle material requires identification of these parameters that are then plugged into the model for failure prediction purposes. The estimation of statistical parameters involves three main steps:

– data collection, i.e. the process of measuring the strength data;

– fractographic examination of failed specimens in order to identify the fracture origins;

– analysis of data sets.

Simple tests such as tensile and flexural tests whose output can be easily interpreted are generally selected. However, there is no limitation in the experimental conditions provided that the stress state can be properly determined as well as the resulting failure probability. Tests on components can even be used, with finite element analysis of stresses and computer code for computation of failure probability.

The statistical parameters can be estimated either from a set of fracture data using the maximum likelihood estimation method or the method of moments, or from a distribution of strength data using identification methods. The latter approach may involve an estimator for the calculation of the probability associated with observed data. The bias of estimator must be measured, i.e. the difference between this estimator's expected value and the true value of the parameter being estimated may need to be corrected. Moreover, after fitting data with one of the models, the goodness of fit must be evaluated. In the presence of multiple populations of flaws, the data pertinent to each population must be properly analyzed.

7.2. Methods of determination of statistical parameters

Whatever the geometry of the specimen and the stress state, it is demonstrated that the equation of failure probability [4.54] reduces to the following expression, when the distribution of elemental strengths is described by power law with constants m and λ_0:

$$P = 1 - \exp - \left[-\frac{V}{V_0} K \left(\frac{\sigma_{ref}}{\lambda_0} \right)^m \right] \qquad [7.1]$$

where σ_{ref} is the reference stress (peak stress) in the stressed volume V. λ_0 is a scale factor: σ_0 is the Weibull model, S_0 is the multiaxial elemental strength model. m is the shape parameter. K is obtained by integrating the stress state over the stressed volume V. K depends on the probabilistic model that is considered. It can be demonstrated that the Weibull equation of failure probability is a particular solution of [7.1]: under uniform uniaxial tension, K = 1, σ_{ref} is the specimen tensile strength, such that:

$$P = 1 - \exp\left[-\left(\frac{V}{V_0}\right)\left(\frac{\sigma}{\sigma_0}\right)^m\right] \qquad [7.2]$$

7.2.1. *Maximum likelihood technique*

The parameter estimates obtained using the maximum likelihood technique are unique (for a two-parameter Weibull distribution), and they statistically approach the true values of the population more efficiently than other parameter estimation techniques as the size of the sample increases. Values of m and λ_0 are determined from the set of experimental strength data σ_R^j. They are given by the maximum of the probability density function. For the two-parameter Weibull distribution expressed by equation [7.3], the likelihood function L is defined as expression [7.4]):

$$P = 1 - \exp - \left(\frac{\sigma}{\sigma_v}\right)^m \qquad [7.3]$$

$$L = \underset{j=1}{\overset{n}{\pi}} \left(\frac{m}{\sigma_v}\right)\left(\frac{\sigma_R^j}{\sigma_v}\right)^{m-1} \exp - \left(\frac{\sigma_R^j}{\sigma_v}\right)^m \qquad [7.4]$$

Parameter estimates are determined by taking the partial derivatives of the logarithm of the likelihood function with respect to *m* and σ_v and equating the resulting expressions to zero.

$$\frac{\partial LnL}{\partial m} = 0 \quad \text{for estimation of m} \qquad [7.5]$$

$$\frac{\partial LnL}{\partial \sigma_v} = 0 \quad \text{for estimation of } \sigma_v$$

The system of equations obtained by differentiating the log likelihood function is given by:

$$\frac{\sum_{j=1}^{n} (\sigma_R^j)^m Ln\sigma_R^j}{\sum_{j=1}^{n} (\sigma_R^j)^m} - \frac{1}{n}\sum_{j=1}^{n} ln\left(\sigma_R^j\right) - \frac{1}{m} = 0 \qquad [7.6]$$

$$\sigma_V = \left[\frac{1}{m}\sum_{j=1}^{n}(\sigma_R^j)^m\right]^{1/m} \quad [7.7]$$

Equation [7.6] is solved first for *m*. Subsequently, σ_v is computed from equation [7.7]. The scale parameter λ_0 is derived from σ_v using the following expression obtained by equating expressions [7.1] and [7.3]:

$$\lambda_0 = \sigma_v\left[\frac{V}{V_0}K\right]^{\frac{1}{m}} \quad [7.8]$$

Obtaining a closed-form solution of equation [7.6] for *m* is not possible. This expression must be solved numerically.

7.2.2. *Method of moments*

In mathematics, a moment is a specific quantitative measure of the shape of a set of points. If the points represent probability density, then the zeroth moment is the total probability (i.e. one), the first moment is the mean, the second moment is the variance and the third moment is the skewness. For a two-parameter Weibull distribution (expressed as [7.2]), the mean and the variance are related to statistical parameters by the following expressions:

$$\bar{\sigma} = \frac{\sigma_0 V_0^{1/m}\Gamma(1+1/m)}{V^{1/m}} \quad [7.9]$$

$$\mathrm{Var}(\sigma) = \frac{\sigma_0^2}{V^{\frac{2}{m}}}\left[\Gamma(1+\frac{2}{m}) - \Gamma^2(1+\frac{1}{m})\right] \quad [7.10]$$

where $\Gamma(1 + x) = x\,!$ is the Gamma function, n is an integer.

The coefficient of variation (CV) is a standardized measure of dispersion of a probability. It is defined as the ratio of the standard deviation to the mean:

$$CV = \frac{\sqrt{Var(\sigma)}}{\bar{\sigma}} = \frac{\left\{\Gamma\left(1+\frac{2}{m}\right)-\Gamma^2\left(1+\frac{1}{m}\right)\right\}^{1/2}}{\Gamma\left(1+\frac{1}{m}\right)} \qquad [7.11]$$

Equation [7.11] is solved first for m. Then, σ_0 is calculated from equation [7.9]. The scale factor λ_o is derived from σ_0 using expression [7.12] obtained by equating [7.1] and [7.2]. Introducing formula for σ_0 derived from equation [7.9], such that:

$$\lambda_0 = \bar{\sigma} \frac{K^{1/m} V^{1/m}}{\Gamma(1+\frac{1}{m}) V_o^{1/m}} \qquad [7.12]$$

7.2.3. *Fitting theoretical distribution function to empirical one*

The estimated statistical parameters are such that the values of probability of fracture calculated from equation [7.1] fit the distribution of strengths obtained experimentally. Various fitting techniques can be used, like the method of least squares. The best fit in the least-squares sense minimizes the sum of squared residuals, a residual being the difference between an observed value and the fitted value provided by a model.

The linearized form of equation [7.1] and linear regression analysis can be used:

$$\text{LnLn}\ \frac{1}{1-P} = \text{m Ln}\ \sigma_{ref} + \text{Ln}\left[\frac{V}{V_o}\frac{K}{\lambda_o^m}\right] \qquad [7.13]$$

m and λ_0 are estimated from the slope (A) and intercept (C) of the linear regression equation Y = AX + C + e with Y= LnLn $(\frac{1}{1-P_i})$, X = Ln σ_R^i, where e is the error term.

P_i is the value of failure probability associated with the value of σ_R^i (section 7.3). From the comparison of the best fit line with theoretical expression [7.13], we obtain:

$$m = A$$

$$\mathrm{Ln}\left[\frac{V}{V_o}\cdot\frac{K}{\lambda_o^m}\right] = \mathrm{C} \Rightarrow \lambda_o = \left[\frac{V}{V_o}\cdot\frac{K}{e^c}\right]^{1/m} \qquad [7.14]$$

The data fitting method requires that failure probability values be associated with strength data. This is a disadvantage as discussed in section 7.3. However, an advantage of this method is that it allows checking whether one or several populations of flaw populations are present. A single population is indicated by a single regression line, whereas several regression lines suggest the presence of several flaw populations.

7.2.4. *Fitting failure probability computations to empirical values*

This method can be employed when failure probability is computed using a software. It consists of comparing the empirical value of failure probability to that calculated for an arbitrary value of λ_0, the value of m having been determined by the first or the second method proposed above. Values of λ_0 are obtained using the following equation:

$$\lambda_o = \lambda_{oA}\left[\frac{Ln(1-P_{CERAM})}{Ln(1-P_{EXP})}\right]^{1/m} \qquad [7.15]$$

λ_{oA} is an arbitrary value of λ_o plugged into the computer code for computation of failure probability. This latter is denoted by P_{CERAM} because it refers to the post-processor CERAM used, which will be discussed in Chapter 8. P_{exp} is the empirical failure probability. Both P_{CERAM} and P_{exp} correspond to the same value of stress σ_{ref}.

7.2.5. *Fitting the tensile behavior curve of multifilament bundles*

This technique was developed recently on bundles of parallel brittle fibers such as ceramic and glass fibers. It has been applied successfully to a wide variety of fibers. Parameters of statistical distributions of filament strengths are estimated by fitting the equation of tensile behavior of bundle to experimental curve. The force-stress equation is given by the bundle model of parallel and independent fibers [CAL 04, RMI 12]:

$$F(\sigma) = N_t\, S_f \sigma\, [1 - P(\sigma)] \qquad [7.16]$$

where N_t is the initial number of intact filaments in tow, S_f is the average filament cross-sectional area. $P(\sigma)$ is the failure probability of filament subject to stress σ.

The bundle models are based upon the following hypotheses: (1) the bundle contains identical and parallel fibers (having radius R_f and length l), (2) when a fiber fails, the load is carried equally by all the surviving fibers, whereas the broken fiber no longer carries any load. Consequently, the total force F (σ) applied to the bundle is:

$$F(\sigma) = N_t(1 - \sigma(\sigma))S_t\sigma \qquad [7.17]$$

The fraction of broken fibers $\alpha(\sigma)$ is the ratio of the number of broken fibers N to the initial number N_t. It represents the failure probability of the Nth fiber under stress σ (the fibers being ranked according to ascending order in strength):

$$\alpha(\sigma) = \frac{N}{N} = P(\sigma) \qquad [7.18]$$

The statistical distribution of fiber strengths is described by the Weibull model or by the Normal distribution [RMI 12]. Figure 7.1 shows an example of fitting of force-strain curve by equation [7.16].

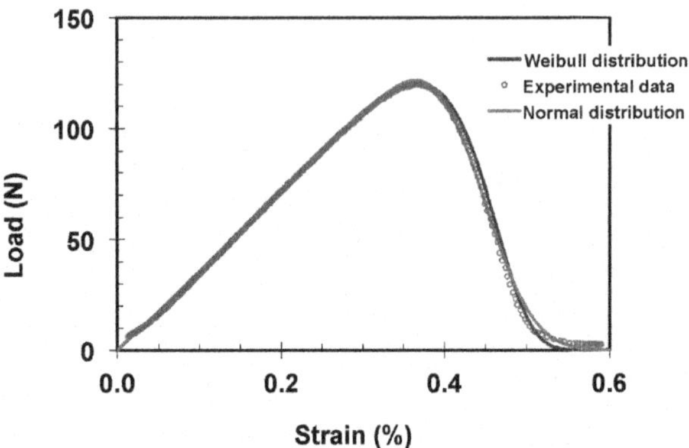

Figure 7.1. *Fitting experimental behavior of SiC tows for Weibull and normal distributions of filament strengths [RMI 12] . For a color version of the figure, see www.iste.co.uk/lamon/brittle.zip*

7.2.6. *Examples*

The methods of determination of statistical parameters are illustrated on the example of 3-point bending for the Weibull and multiaxial elemental strength models:

– For the *Weibull model*: $K_W = \dfrac{1}{2(m+1)^2}$ (see Chapter 6)

– For the *multiaxial elemental strength model*: $K_M = K_W$ if uniaxial stress state is assumed in the specimen (see Chapter 6). When a biaxial stress state is considered:

$$K_M = \frac{1}{VI_V(m,0,0)} \int_V \left(\frac{\sigma_1}{\sigma_{\max}}\right)^m I_V(m, \frac{\sigma_2}{\sigma_1}, 0) dV \qquad [7.19]$$

– *Maximum likelihood technique*: m and λ_0 are given by the following expressions:

m_w : equation [7.6]

$$\sigma_0 = \left[\frac{1}{n}\sum_{j=1}^{n}\left(\sigma_R^j\right)^m\right]^{1/m}\left[\frac{V}{V_o 2(m+1)^2}\right]^{1/m} \qquad [7.20]$$

$$m_M = m_W$$

$$S_0 = \left[\frac{1}{n}\sum_{j=1}^{m}\left(\sigma_R^j\right)^m\right]^{1/m}\left[K_M\right]^{1/m} \qquad [7.21]$$

where m_W and σ_o are the Weibull statistical parameters; m_M and S_o are those statistical parameters of the multiaxial elemental strength model.

– *Method of moments*: $m_M = m_W$ according to equation [7.11]

$$\sigma_0 = \bar{\sigma}\frac{1}{\left[2(m+1)^2\right]^{1/m}}\frac{V^{1/m}}{\Gamma\left(1+\dfrac{1}{m}\right)V_o^{1/m}} \qquad [7.22]$$

$$S_0 = \bar{\sigma}K_M^{1/m}\frac{V^{1/m}}{\Gamma\left(1+\dfrac{1}{m}\right)V_o^{1/m}} \qquad [7.23]$$

– *Fitting linearized equation distribution function to the empirical one*:

$$m_W = m_M$$

$$\sigma_0 = \left[\frac{V}{V_o}\frac{e^{-c}}{2(m+1)^2}\right]^{1/m} \qquad [7.24]$$

$$S_0 = \left[\frac{Ve^{-c}}{V_o}K_M\right]^{1/m} \qquad [7.25]$$

7.3. Production of empirical data

The fracture stresses σ_{Ri} represent the experimental values of reference stresses σ_{ref}. They are measured by means of simple tests on specimens having simple shapes, or, possibly, on components of complicated shapes subjected to complex stress states calculated using numerical analysis. Tensile tests on massive brittle materials are not recommended due to practical difficulties associated with grip failures and misalignment of the specimen with respect to loading direction. In such conditions, the current stress state in the test specimen differs from theoretical uniform stress state. Bending tests are widely used. Tensile tests are used on fibers.

The stresses are derived from the forces measured during mechanical tests, using closed-form expressions for few simple tests including traction, bending, etc. The failure stress is derived from the value at failure of the reference (peak) stress in the specimen or in the component.

Closed-form expressions of σ_R^j as functions of force measured at failure using tensile and bending tests are given below:

$$\text{traction: } \sigma_R^j = \frac{F_R^j}{S_j} \qquad [7.26]$$

where S_j is the cross-sectional area of the ith specimen.

$$\text{3-point bending: } \sigma_R^j = \frac{3F_R^j l}{4bd^2} \qquad [7.27]$$

$$\text{4-point bending: } \sigma_R^j = \frac{3F_R^j l_2}{4bd^2} \qquad [7.28]$$

where 2l is the span length, l_2 is the distance between a loading point and the nearest support, 2d is the height and b is the width of test specimen (see Chapter 5).

For the determination of statistical parameters using the maximum likelihood technique, the σ_R^j values are introduced into appropriate equations given in section 1.1. Note that σ_0 and S_0 are determined from equation [7.8].

With the method of moments, the mean and standard deviation of σ_R^j values are calculated first. The statistical parameters are determined then using equations [7.9], [7.11] and [7.12].

With the method of fitting theoretical and empirical cumulative distribution functions, values of failure probability must be associated with the experimental σ_R^j data. For this purpose, the strength data σ_R^j are ranked in ascending order:

$$\sigma_R^1 < \sigma_R^2 < \sigma_R^3 < \ldots < \sigma_R^i < \ldots < \sigma_R^n,$$

where the i is the rank and n is the number of data.

Then, the ith result is assigned a ranked probability of failure P_i. The functions used to calculate these probabilities are called estimators. Estimators have the following form: $P_i = \dfrac{i-\alpha}{n+\beta}$

Historically, the values of constants α and β have been selected to minimize the bias on the Weibull modulus estimated $(\hat{m})$ on a limited sample of n testspecimens. They are related through the following equation: $\beta = 1 - 2\alpha$ $(0 \leq \alpha \leq 1)$.

The estimators that are employed are:

$$P_i = \frac{i}{n+1} \qquad [7.29]$$

$$P_i = \frac{i-0.3}{n+0.4} \qquad [7.30]$$

$$P_i = \frac{i-3/8}{n+1/4} \quad [7.31]$$

$$P_i = \frac{i-0.5}{n} \quad [7.32]$$

Barnett [BAR 75] discussed the choice of α. More recently, Gong [GON 00] proposed particular values for α and β that are independent: $\alpha = 0.999$ and $\beta = 1000$. He showed that these values yielded precise estimates of m and σ_o.

7.4. Bias and variability

There has been a great deal of papers on the accuracy and variability of estimated Weibull statistical parameters. The estimates of statistical parameters depend on the method of estimation chosen, on the probability estimator for construction of Weibull plot, the sample size and the quality of fit of the calculated probability function to the empirical distribution. They show variation that depends on sampling and sample size. Typically, the scatter increases with decreasing sample size. The estimation of Weibull statistical parameters from experimental failure data may be skewed for the following reasons:

– the use of biased estimators for the determination of failure probabilities associated with experimental data;

– the use of undersized sample, especially for highly heterogeneous materials, i.e. containing large amounts of flaws with broad size range;

– the uncertainty in strength data calculated from measured forces, which may be significant for certain geometries (like small diameter fibers) or loading conditions.

Authors look for methods of correction of estimates obtained on limited sample sizes. For this purpose, they use more or less complex analyses and computations to define appropriate estimators of experimental failure probability or they introduce additional parameters into the Weibull equation in order to improve the fit to experimental distribution of strength data.

7.4.1. *Bias of estimators and methods of estimation*

Most of the estimators are biased. In several papers [BER 84, DAV 04, GRI 03], the accuracy of probability estimators has been investigated using computer-generated "experimental" data. The Monte-Carlo method was used to generate random samples of various numbers of specimens, assuming that the failure stresses have a Weibull distribution. Failure stress values σ_R^i were calculated for known m and σ_0 as $\sigma_R^i = \sigma_0 Ln\left(\frac{1}{1-X_i}\right)^{1/m}$, where X_i is a random variable with uniform distribution between 0 and 1. By comparing the calculated modulus from different estimators with the known modulus, a set of correction factors was determined for each estimator. The correction factors were found to depend on the number of samples in the set and on the estimator.

A similar investigation was conducted using actual experimental results obtained on a set of 100 bars of zirconia [SUL 86]. The statistical parameters pertinent to actual ceramic are not known, and whether the failure stresses have a Weibull distribution is also unknown. The Weibull modulus was calculated for various sets by least squares fitting. Estimates for various estimators and various sample sizes are compared in Figure 7.2. Estimator $P_i = \frac{i-0.5}{n}$ yields estimates of Weibull modulus that are not significantly dependent on sample size. The estimator $P = \frac{i}{n+1}$ yields lower values of Weibull modulus. Intermediate results were obtained with the two other estimators (equations [7.30]–[7.31]). The results indicate that the use of a large sample size does not satisfactorily compensate for the use of a poor estimator.

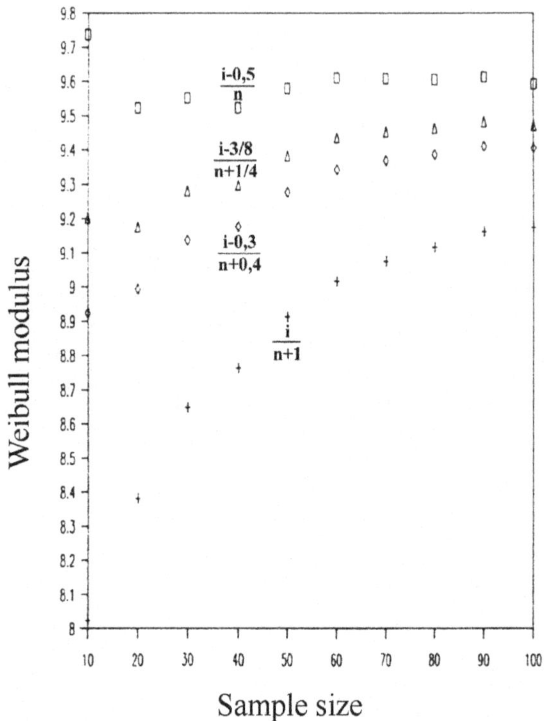

Figure 7.2. *Influence of probability estimator and sample size on estimates of Weibul modulus [SUL 86, LAU 82]*

Table 7.1 compares the methods of estimation [TRU 79]. Random samples of various sizes were simulated from the Weibull distribution for m = 5. The estimates were calculated using the least squares method (m_S) and the linear regression method (m_B) together with the best and the poor previous estimators, the method of maximum likelihood (m_L) and the method of moments (m_M). Table 7.1 shows the mean values of correction factors m_S/m, m_B/m, m_L/m and m_M/m, and the estimated errors of the distributions of correction factors. For an unbiased estimator, the mean value of correction factor is close to 1. On this basis, it appears that the use of least square method with a sample size of about 40 and estimator $P_i = \dfrac{i-0.5}{n}$ yields the best estimate of m. Table 7.1 can be used to adjust an estimate for its bias. For example, if m_L = 11.5 and n = 40, then m should be estimated by 11.5/1.035 = 11.1 with standard error 11.1 × 0.13 = 1.4.

n	m_S/m $P_f = \frac{i}{n+1}$	m_S/m $P_f = \frac{i-\frac{1}{2}}{n}$	m_B/m $P_f = \frac{i}{n+1}$	m_B/m $P_f = \frac{i-\frac{1}{2}}{n}$	m_L/m	m_M/m
10	0.867	1.055	0.944	1.129	1.165	1.102
	(0.28)	(0.34)	(0.30)	(0.36)	(0.34)	(0.35)
20	0.894	1.0109	0.952	1.063	1.078	1.049
	(0.20)	(0.23)	(0.20)	(0.23)	(0.20)	(0.21)
30	0.910	1.008	0.961	1.039	1.048	1.031
	(0.17)	(0.19)	(0.16)	(0.17)	(0.16)	(0.17)
40	0.920	1.002	0.964	1.028	1.035	1.021
	(0.15)	(0.16)	0.14)	(0.15)	(0.13)	(0.14)
50	0.924	0.996	0.966	1.020	1.025	1.014
	(0.14)	(0.14)	(0.12)	(0.13)	(0.12)	(0.12)

Table 7.1. *Estimated means of m obtained for various sample sizes and methods: least squares (m_S), linear regression (m_B), maximum likelihood (m_L), moments (m_M). Standard errors are given in [TRU 79]*

Tables 7.2–7.5 compare the method of linear regression for estimator $P_i = \frac{i-0.5}{n}$. and the method of maximum likelihood [KHA 91, BER 84] using computed generated failure data and given statistical parameters m and σ_0. The best estimate for m is obtained with estimator $P_i = \frac{i-0.5}{n}$ and 50 specimens. Comparable estimate of m with the maximum likelihood method requires a larger sample size. However, smaller coefficients of variation were obtained with the maximum likelihood method. The estimate of scale factor is neither affected by the method nor by the number of failure stress data. Tables 7.2–7.5 also give the values of correction factors for adjusting an estimate for its bias. The unbiased values of m and σ_0 are obtained by multiplying the correction factor by the estimate:

$$m = k_m \, \hat{m} \tag{7.33}$$

$$\sigma_0 = k_{\sigma_0} \hat{\sigma}_0 \tag{7.34}$$

where $\hat{m}$ and $\hat{\sigma}_0$ are the estimates of m and σ_0, and k_m and k_{σ_0} are the correction factors.

Sample size n	$\hat{m}$	sd_m	CV	$\frac{\hat{m}}{m} = \frac{1}{k_m}$
3	8.802	13.14	1.49	1.760
4	6.459	4.249	0.659	1.291
5	5.994	3.442	0.574	1.198
6	5.682	2.588	0.455	1.136
7	5.555	2.400	0.432	1.111
8	5.476	2.040	0.372	1.095
9	5.320	1.873	0.353	1.064
10	5.310	1.732	0.326	1.062
12	5.216	1.528	0.293	1.043
14	5.190	1.425	0.274	1.038
16	5.152	1.295	0.244	1.027
18	5.108	1.220	0.238	1.021
20	5.077	1.124	0.221	1.015
22	5.080	1.109	0.209	1.017
24	5.077	1.004	0.1979	1.015
26	5.066	0.939	0.185	1.012
28	5.035	0.950	0.188	1.007
30	5.031	0.931	0.185	1.006
35	5.032	0.897	0.178	1.006
40	5.012	0.807	0.161	1.002
50	4.998	0.728	0.145	0.999

Table 7.2. *Estimated means of m determined using linear regression analysis and probability estimator $P_i = \frac{i-0.5}{n}$ for various sample sizes. sd_m = standard deviation, CV = coefficient of variation, $\hat{m}$ is the mean of 5,000 simulations [KHA 91, BER 84]*

Sample size n	$\hat{m}$	sd_m	CV	$\frac{1}{k_m}=\frac{\hat{m}}{m}$
3	7.935	5.111	0.644	1.587
4	7.422	4.072	0.548	1.484
5	6.912	3.339	0.483	1.382
6	6.553	2.808	0.422	1.310
7	6.334	2.431	0.383	1.266
8	6.122	2.106	0.344	1.224
9	5.930	1.889	0.318	1.186
10	5.825	1.763	0.302	1.165
12	5.667	1.467	0.258	1.133
14	5.600	1.295	0.231	1.120
16	5.479	1.201	0.219	1.095
18	5.461	1.132	0.207	1.092
20	5.355	1.043	0.194	1.071
22	5.343	0.973	0.182	1.068
24	5.335	0.917	0.171	1.067
26	5.293	0.879	0.166	1.058
28	5.276	0.813	0.154	1.051
30	5.244	0.794	0.151	1.048
35	5.201	0.713	0.137	1.040
40	5.189	0.672	0.129	1.037
45	5.165	0.608	0.1177	1.033
50	5.130	0.561	0.109	1.026

Table 7.3. *Estimated means of m determined using maximum likelihood for various sample sizes. sd_m = standard deviation, CV = coefficient of variation, m is the mean of 5,000 simulations [KHA 91, BER 84]*

Sample Size n	$\hat{\sigma}_0$	$sd\sigma_0$	$CV\sigma_0$	$\frac{\hat{\sigma}_0}{\sigma_0}=\frac{1}{k_{\sigma_0}}$
3	1.076	0.134	0.125	0.988
5	1.089	0.114	0.105	0.994
7	1.087	0.088	0.081	0.998
10	1.090	0.072	0.063	1.001
20	1.089	0.052	0.047	1.000
30	1.091	0.043	0.039	1.001
40	1.090	0.040	0.036	1.000
50	1.0914	0.032	0.030	1.001

Table 7.4. *Estimated means of scale factor determined using linear regression analysis and probability estimator $P_i=\frac{i-0.5}{n}$ for various sample sizes. $sd\sigma_0$ = standard deviation, $CV\sigma_0=\frac{sd\sigma_0}{\hat{\sigma}_0}$ = coefficient of variation, $\hat{\sigma}_0$ is the mean of 5,000 simulations [KHA 91, BER 84]*

Sample Size n	$\hat{\sigma}_0$	$sd\sigma_0$	$CV\sigma_0$	$\frac{1}{k_{\sigma_0}} = \frac{\hat{\sigma}_0}{\sigma}$
3	1.063	0.129	0.121	0.976
5	1.070	0.113	0.105	0.982
7	1.078	0.087	0.081	0.991
10	1.083	0.074	0.068	0.994
20	1.085	0.050	0.046	0.996
30	1.087	0.040	0.037	0.998
40	1.087	0.036	0.033	0.997
50	1.087	0.031	0.029	0.998

Table 7.5. *Estimated means of scale factor determined using maximum likelihood for various sample sizes. sd* σ_o *= standard deviation,* $CV\sigma_0 = \frac{sd\sigma_o}{\hat{\sigma}_o}$ *= coefficient of variation ,* $\hat{\sigma}_0$ *is the mean of 5,000 simulations [KHA 91, BER 84]*

Lissart [LIS 97a] compared the method of linear regression for estimator $P_i = \dfrac{i-0.5}{n}$ and the method of maximum likelihood on experimental failure data obtained on SiC Nicalon fibers (batches of 30 test specimens). The best fit of the calculated and empirical distribution functions was obtained with the values of statistical parameters estimated using the maximum likelihood method. However, it was also shown that the linear regression method is capable of providing satisfactory estimates of statistical parameters.

The above correction factors can be applied to the estimates of statistical parameters that appear in equations of failure probability based on the power law flaw strength density function. According to equation [7.1], the shape parameter is given by m, and the scale factor $\sigma_0 = \dfrac{\lambda_0}{\left(\dfrac{V}{V_0}K\right)^{1/m}}$. Unbiased values of m and λ_0 can be derived from estimates using equation [7.33] and $\lambda_0 = k_{\sigma 0}\,\hat{\lambda}_0$ [7.35].

7.4.2. *Variability of statistical parameters*

Investigation of variability of statistical parameters requires a huge amount of experimental failure stresses, which is generally difficult to produce on ceramic test specimens for practical reasons of time and cost. Large sets of failure stresses can be generated using single tensile tests on multifilament tows that contain several hundreds or thousands filaments. This technique was applied to glass or ceramic tows with 500–2,000 filaments [RMIL 12]. Statistical distributions of filament strengths were determined by fitting equation [7.16] to tow experimental tensile curve (Figure 7.1). The statistical distributions of filament strengths were found to follow normal as well as Weibull distributions (Figure 7.3). A perfect agreement between both distributions was obtained (Figures 7.1 and 7.3), and a quite negligible scatter in statistical parameters was observed. Thus, estimates of Weibull modulus m = 5.2–5.4 were obtained for SiC fibers as opposed to the wide variability that is reported in the literature (m = 2–8). It was concluded that flaw strengths are distributed normally and that the statistical parameters that were derived are the true ones. By contrast, the conventional method of estimation of Weibull parameters with estimator $P_j = (j - 0.5)/n$ gave estimates of m showing more significant variation (m = 4.8–5.2) (Table 7.6).

	Number of filaments	Normal distribution Curve fitting		Weibull distribution Curve fitting		Weibull plot estimator	
		$\bar{\varepsilon}$ (%)	sd (%)	m	ε_l (%)	m	ε_l (%)
Tow 1	487	1.16	0.25	5.30	1.25	5.23	1.26
Tow 2	986	1.11	0.24	5.23	1.20	4.86	1.20
Tow 3	924	1.15	0.24	5.43	1.23	5.12	1.23
Subsets 1	20 (5)					4.7–7.2	1.20–1.31
Subsets 2	30 (5)					4.6–6.1	1.23–1.24
Subset 3	168 (1)					11.2	0.8

Table 7.6. *Estimates of statistical parameters of Normal distributions ($\bar{\varepsilon}$ = mean strain, sd = standard deviation) and Weibull distributions (ε_l = characteristic strain for l = 115 mm) of failure strains for SiC Nicalon filaments [RMI 12]. The number of subsets is given in brackets*

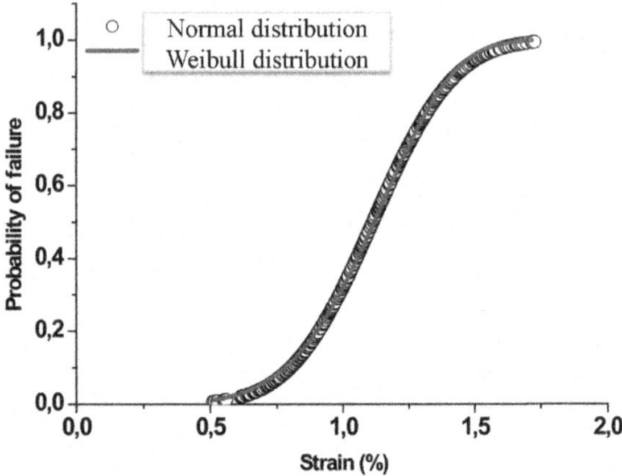

Figure 7.3. *Cumulative Normal and Weibull distribution functions of failure strains for SiC Nicalon filaments derived from tensile behavior of multifilament tows [RMI 12]. For a color version of the figure, see www.iste.co.uk/lamon/brittle.zip*

A significantly greater scatter was obtained with smaller size sets of data selected randomly among one of the sets of 1,000 data: samples of 20 (m = 4.7–7.2), 30 (m = 4.5–6.1), 168 (m = 11.2). The subset of 168 data comprised only the lowest strengths derived from the failures prior to maximum load (Figure 7.1). Five sets of 20 or 30 samples and one set of 168 samples had been selected. These numbers of sets are very small when compared to the amount of data sets that can be extracted from the original one, as given by the binomial coefficient:

$$C_n^N - \frac{N!}{(N-n)!n!} \qquad [7.36]$$

For n = 20 and n = 30, the number of possible data sets is quite huge: $C_{20}^{1000} = 3.4.10^{41}, C_{30}^{1000} = 3.1.10^{27}$.

This indicates how high the probability that authors used significantly different samples for the estimation of Weibull parameters of SiC fibers is. These results demonstrate that the selection of samples is responsible for variability in statistical parameters.

7.4.3. *Goodness of fit*

To check the goodness of estimated statistical parameters, we should evaluate the goodness of fit of calculated distribution to the experimental failure results. A statistic measuring the difference between calculated values of a given distribution function and corresponding values of empirical distribution function (EDF) is called an EDF statistic. There are several possible approaches, some graphical and some numerical. Our intend is not to discuss EDF statistics in detail – this topic is covered in [DAG 86] and [LAW 82] – but instead to indicate techniques that may be probably the best for evaluating the potential quality of failure predictions from the estimates of statistical parameters.

The Chi-squared test is found in almost all textbooks. It is simple to apply, and tables of critical values are readily available. However, it requires more than 70 data inputs to able to reject inappropriate distribution.

The Kolmogorov–Smirnov test is considered as probably the best known of the EDF tests [SNE 88]. It measures discrepancies between successive values of the EDF i/n at x_i, and corresponding calculated values $F(x_i)$ of the distribution function. The Kolmogorov–Smirnov statistic measures the maximum discrepancy:

$$D_n^+ = \max\left\{\frac{i}{n} - F(x_i)\right\} \quad [7.37]$$

$$D_n^- = \max\left\{F(x_i) - \frac{i-1}{n}\right\} \quad [7.38]$$

$$D_n = \max\left\{D_n^+ \; D_n^-\right\} \quad [7.39]$$

The EDF Cramer–von Mises test is better than the Kolmogorov–Smirnov test since it deals with several points rather than a single point:

$$\omega_n^2 = \sum_{i=1}^{n}\left\{F(x_i)-\frac{i-0.5}{n}\right\}^2+\frac{1}{(12n)} \quad [7.40]$$

According to Snedden [SNE 88], the Anderson–Darling test is the most powerful of the EDF tests [AND 52]. This test is based on the area of discrepancy between the empirical and the calculated values of distribution functions. It gives more importance to the tails of the distribution, and is the most appropriate for small samples. The test statistic is:

$$A_n^2 = -\text{n} - \sum_{1}^{n}\frac{2i-1}{n}\left[ln F(x_i)+ln(1-F(x_{n+1-i}))\right] \quad [7.41]$$

Snedden [SNE 88] proposes the following approximate rule of thumb: only values of A_n^2 < 0.5 are likely to have meaningful significance.

As an illustrative example, it is worth pointing out that the use of the maximum likelihood estimation method coupled with the Anderson–Darling test yielded statistical parameters for SiC fibers that were similar to the values estimated by fitting the tensile behavior curve of multifilament bundles (sections 7.2.5 and 7.4.2) [LIS 97a, RMI 12].

7.5. Effect of the presence of multimodal flaw populations

Several populations of fracture-inducing flaws may be present in the test specimens, which may be reflected by the plot of lnln(1/1-P_i) vs lnσ_R^i that would consist of distinct parts (see Chapter 1). However, such diagram shape does not provide reliable evidence of the presence of multimodal populations of flaws. It may result from both an insufficient number of data and the effect of probability estimator that exaggerates the weight of strength extremes. An example of such distortion is provided in Figure 7.4. By contrast, it can be accepted that a single domain diagram characterizes a single population of flaws even if these flaws show differences in terms of location in the specimen (surface vs. volume), shape and type.

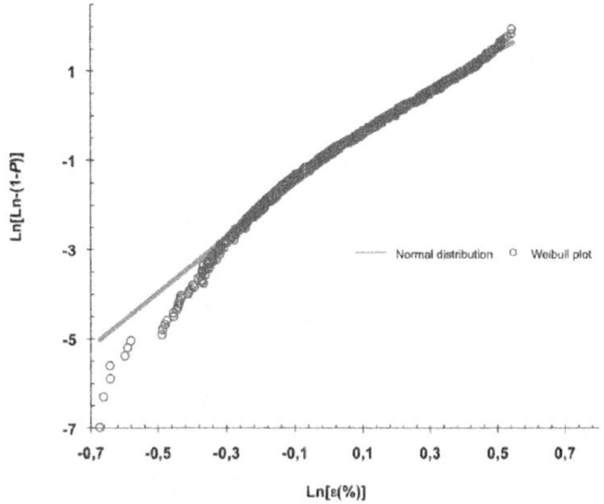

Figure 7.4. *Comparison of Weibull plot of strains-to-failure of SiC Nicalon filaments (probability estimator j/N) with the Normal distribution derived from the tensile behavior of multifilament tows. For a color version of the figure, see www.iste.co.uk/lamon/brittle.zip*

Fractographic examination of each failed specimen is recommended in order to characterize the fracture origins. Three situations may arise:

– the populations of defects are exclusive: the defects of a family are present in certain specimens, those belonging to the other family (case of two families) are present in the rest of specimens. The definition can be extended easily when the number of families is > 2;

– the populations of defects are concurrent: the defects of every family are present in all the samples;

– the populations of defects are partially concurrent: the defects of a population are present in all the samples, while those of the second family are present only in some of them.

As an illustrative example, below are given the equations of failure probability for two families which are referred to as populations A and B.

7.5.1. *Exclusive populations*

Each test specimen contains defects of one population (A or B) only. The failure stresses corresponding to each single population are handled separately and independently using the methods presented above.

7.5.2. *Concurrent populations*

The defects of both families A and B are present in every test specimen. The fracture of a test specimen results from a defect of family A or B. The probability of failure of a test specimen is thus given by the following equation, assuming that the populations of defects are independent:

$$P = 1 - (1 - P_A)(1 - P_B) \qquad [7.42]$$

where P_A and P_B are the probabilities of failure, respectively, from the defects of the family A and the defects of the family B.

Equation [7.42] was established as follows. Assuming that every defect is isolated in an element of volume that is independent of others, we can represent the material as a chain of volume elements, a number of which contains defects of the family A (number n_A) whereas the others (number n_B) contain defects of the family B. According to the weakest link concept, the survival of the chain requires the survival of all the elements. Thus, equation [7.42] becomes:

$$1 - P = (1 - P(a_1, V_A))^{n_A} \; (1 - P(a_2, V_B))^{n_B} \qquad [7.43]$$

where $P(a_1, V_A)$ is the probability of failure from a defect of size a_1, the family A, V_A being the elementary volume for the defects of the family A. Also, $P(a_2, V_B)$ is the probability of failure from a defect of size a_2, the family B, V_B being the corresponding elementary volume.

According to equation [7.1], it comes:

$$P_A = 1 - \exp\left[-\frac{V}{V_o} K_A \left(\frac{\sigma_{ref}}{\sigma_{oA}}\right)^{m_A}\right] \quad [7.44]$$

$$P_B = 1 - \exp\left[-\frac{V}{V_o} K_B \left(\frac{\sigma_{ref}}{\sigma_{oB}}\right)^{m_B}\right] \quad [7.45]$$

Substituting expressions [7.44] and [7.45] for P_A and P_B in [7.42], we obtain:

$$P = 1 - \exp\left[-\frac{V}{V_0}\left[K_A\left(\frac{\sigma_{ref}}{\sigma_{0A}}\right)^{m_A} + K_B\left(\frac{\sigma_{ref}}{\sigma_{0B}}\right)^{m_B}\right]\right] \quad [7.46]$$

It is necessary to determine two couples of parameters (m_A, σ_{oA}) and (m_B, σ_{oB}). We can use an iterative method of identification of couples of data such as the equation [7.46] fits the empirical distribution function. An alternative simple method consists of separating the failure stress data according to fracture origin, by taking into account the presence of the other population [JOH 83]. Thus, methods of determination of statistical parameters discussed in previous sections are applied to each set of failure data. This approach is described in subsequent section 7.5.4.

7.5.3. *Partially concurrent populations*

The defects of the family A are present in all the test specimens, whereas the defects of the family B are present in a few test specimens only. The expression for the probability of fracture is:

$$P = (1 - \pi)\, P_B + \pi\, (1 - (1 - P_A)\,(1 - P_B)) \quad [7.47]$$

π is the fraction of test specimens containing both concurrent populations A and B.

The value of π is derived from the result of fractographic analysis. Then, the failure stresses can be seperated into two subsets: on the one hand, the subset corresponding to failures from defects of population A, and on the other hand that corresponding to failures from defects of both populations A and B. The two subsets of data correspond to exclusive flaw populations: respectively, A on the one hand and A + B on the other hand. They can be handled separately: the first one as a single population of data, the second one as concurrent populations of data.

7.5.4. *Concurrent populations of defects: separation of data*

The censored data method proposed by Johnson for separation of the data means reranking the data of every family with regard to the data of the other families. The new rank is calculated for the data of each population using the following formulas [JOH 83]. A new value of increment Δ is calculated as soon as one or more censored data of the other population are encountered in the sequence of test data.

$$i'(i) = i'(i-1) + \Delta \qquad [7.48]$$

$$\Delta = \frac{n+1-i'(i-1)}{n+2-j} \qquad [7.49]$$

i refers to the rank of a datum in the considered distribution (data are in ascending order), i'(i) is the new rank of the ith failure stress, i'(i–1) is the new rank of the previous datum. j is the former rank of the failure datum in the total distribution having n data.

If the statistical parameters are then estimated by the method of fitting the distribution function to the empirical distribution so obtained, the probability estimator is:

$$P_{i'} = \frac{i' - 0.5}{n} \qquad [7.50]$$

where n is the total number of data in the total distribution.

As an illustrative example of the method of separation of data, let us consider two concurrent populations A and B of data. The total number of data is 10. Table 7.7 gives the initial ranking of all the data and the new ranks for the data of each population calculated using formulas [7.48] and [7.49]. It can be noted that the sum of last new rank and last increment values is 11 for both populations A or B. This is demonstrated as follows: i being the rank of the last datum of a population, for i + 1, j = n + 1, i'(i + 1) = i'(i) + Δ, Δ = n + 1 – i' (i), which gives i'(i + 1) = n + 1. The condition i'(i + 1) = n + 1, with i the last datum of the examined distribution, indicates that the separation of data is valid.

		Population A		Population B	
	j	Δ	i' (i)	Δ	i' (i)
A	1	1	1		
A	2	1	2		
B	3			1.22	1.22
B	4			1.29	2.51
A	5	1.29	3.29		
B	6			1.41	3.92
A	7	1.66	4.95		
A	8	1.99	6.94		
B	9			2.36	6.28
A	10	2.03	8.97		
		2.03	11	4.72	11

Table 7.7. *Example of application of the censored data method for the separation of failure data in the presence of two concurrent flaw populations A and B; the failure data are in ascending order*

7.5.5. *Concurrent populations of defects: maximum likelihood method*

The alternative method of estimation of statistical parameters in the presence of concurrent flaw populations is based on the following likelihood function for the two-parameter Weibull distribution [NEL 82]:

$$L = \left\{ \prod_{i=1}^{r} \left(\frac{m}{\sigma_o}\right)\left(\frac{\sigma_R^i}{\sigma_o}\right)^{m-1} exp-\left(\frac{\sigma_R^i}{\sigma_o}\right)^m \right\}\left\{ \prod_{j=r+1}^{N} exp-\left(\frac{\sigma_R^j}{\sigma_o}\right)^m \right\} \quad [7.51]$$

For the purpose of discussion here, the different flaw populations are referred to as A, B, C, etc. For estimation of statistical parameters for the failure data corresponding to the flaw population A, the first product is carried out for the number (r) of specimens failing from the flaw population A: i is the associated index in the first product. The second product is carried out for the failure data corresponding to the other flaw populations: j is the associated index. As shown earlier for a single flaw population, the estimates of statistical parameters are determined by taking the partial derivatives of the logarithm of the likelihood function with respect to m and σ_0 and equating the resulting expressions to zero.

$$\frac{\sum_{i=1}^{N}\left(\sigma_R^i\right)^m Ln\sigma_R^i}{\sum_{i=1}^{N}\left(\sigma_R^i\right)^m} - \frac{1}{r}\sum_{i=1}^{r} Ln\left(\sigma_R^i\right) - \frac{1}{m} = 0 \quad [7.52]$$

$$\sigma_o = \left[\frac{1}{r}\sum_{i=1}^{N}\left(\sigma_R^i\right)^m\right]^{1/m} \quad [7.53]$$

Equation [7.46] is solved numerically first for m. Obtaining a closed-form solution of equation [7.46] for m is not possible. Subsequently, σ_0 is computed using equation [7.53].

7.6. Fractographic analysis and flaw populations

The fractographic analysis of specimen fracture surfaces allows us to identify the defects responsible for fracture. These can be classified according to several criteria:

– the type: pores, inclusions, cracks, etc. (see Chapter 1);

– the origin: intrinsic and extrinsic;

– the location: surface, volume, angles, etc.

Some fracture origins cannot be identified in a clear way. The ISO International Standards proposes four options for handling the unidentified fracture origins:

1) Assign them to a previously identified fracture origin on the basis of available objective fractographic information.

2) Assign them to that flaw population corresponding to the closest failure data.

3) Assign them to a new population that is treated as a separate population in the censored data analysis.

4) Remove them from the sample.

This last option is not recommended. The choice of one of the solutions will be guided by the shape of the corresponding distribution function. If the presence of unidentified failures is reflected by a segment on the loglog Weibull plot, the failure data should be treated as a concurrent population. If not, it will be logically assimilated to the preponderant population giving its general shape to the distribution function.

7.7. Examples

We cannot summarize unique reference statistical parameters for the ceramic and fragile materials. One reason is that all the data are not available, and that most of the data have been obtained using numerous conditions that are not properly documented. Furthermore, there are various nuances of ceramics with populations of defects depending on processing mode and conditions. We can, however, identify valuable classes of statistical parameters: for fibers, m is generally between 3 and 6. For ceramic and fragile materials, m is between 10 and 25. The scale factor can vary from 10 MPa (internal defects) to 400 MPa (surface-located defects).

Table 7.8 shows the statistical parameters determined for a silicon nitride ceramic. The failure stresses were measured using various flexural tests. The statistical parameters of Weibull and multiaxial elemental strength models were estimated using the various methods presented in this chapter. It can be noted that the smallest variability was observed on those parameters estimated by means of the Finite Element Analysis (FEA) post-processor CERAM for the multiaxial elemental strength model. It can also be observed that the scale factor of the multiaxial elemental strength model is generally smaller than that of Weibull model.

Methods		Surface				Volume
		3 pts (1)	3 pts (2)	3 pts (3)	4 pts	
Linear Regression	σ_o (MPa)	359	332	319	341	244
	S_o (MPa)	330	303	291	310	215
Maximum Likelihood	σ_o (MPa)	291	249	235	283	207
	S_o (MPa)	268	227	215	257	182
Moments	σ_o (MPa)	292	252	238	290	209
	S_o (MPa)	269	230	217	263	184
CERAM	σ_o (MPa)	332	308	298	326	235
	S_o (MPa)	305	282	275	297	208
m	Total	11.2	12.9	11.8	6.4	
	population pores	8.6	8.0	8.0	7.5	11.3

$V_o = 1\ m^3$, $A_o = 1\ m^2$

Table 7.8. *Estimates of statistical parameters for a silicon nitride ceramic obtained using various methods. The failure data were generated using various flexural tests: long span 3-point bending (1), intermediate span 3-point bending (2), short span 3-point bending (3), 4-point bending, σ_o = Weibull scale factor, S_O = scale factor of multiaxial elemental strength model*

Table 7.9 shows the statistical parameters of various ceramics, estimated by means of the software CERAM from failure stresses measured using various test methods.

	Modulus m			Scale factor (MPa)	
	Total population	Surface	Volume	Surface	Volume
Silicon nitride					
3-point bending	9.8		9.8		$\sigma_o = 97$ $S_o = 84$
4-point bending	8.8	9.9	8.7		$\sigma_o = 87$ $S_o = 74.5$
biaxial flexure of discs	8.4	7.9	8.7		$\sigma_o = 80$ $S_o = 72.2$
Al2O3[1]					
4-point bending	23.8	23.8		$\sigma_o = 242$ $\sigma_o = 211$ $S_o = 205$	
3-point bending	23.4	23.4		$\sigma_o = 212.5$ $S_o = 207$	
biaxial flexure of discs	22	22		$\sigma_o = 218$ $S_o = 208$	
Silicon carbide[2]					
4-point bending	7.9		7.9		$\sigma_o = 122$ $\sigma_o = 29$ $S_o = 25$
compression of rings	8.0		8.0		$\sigma_o = 101$ $S_o = 24$

$V_o = 1\ m^3$. $A_o = 1\ m^2$

[1][SHE 83].

[2][FER 86].

Table 7.9. *Estimated statistical parameters for various ceramics. using various testing methods and the FEA post-processor CERAM. The statistical parameters from the literature had been estimated by linear regression analysis of failure data*

Table 7.10 presents the statistical parameters obtained for silicon carbide fibers (Nicalon NL 202). considering that failure was caused by either a unimodal population of volume-located defects. or concurrent populations of volume-located and surface-located defects. or partially concurrent populations of intrinsic defects (volume) and extrinsic defects. The values of index A* (test of Anderson–Darling) indicate that the fitting of caculated to empirical distribution functions is of good quality.

	m	σ_o (MPa)	A*		
Linear Regression	4.91	14.46	0.33	Volume	Single Population
Maximum Likelihood	5.14	17.77	0.31	Volume	
Linear Regression	4.45 6.03	151 47	0.32 0.32	Surface Volume	Concurrent Populations
Maximum Likelihood	4.3 5.1	152 19	0.3 0.3	Surface Volume	
Linear Regression	4.91 2.14	14.49 0.03	0.33 0.42	Intrinsic extrinsic	Partially concurrent Populations
Maximum Likelihood	5.14 3.45	17.73 0.38	0.30 0.34	Intrinsic extrinsic	
Fitting tensile Behavior of tows	5.3	22			Total Distribution

$V_o = 1\ m^3$. $A_o = 1\ m^2$

Table 7.10. *Estimated statistical parameters for SiC fibers (Nicalon) from failure stresses measured using tensile tests on single filaments (gauge length 50 mm) [LIS 97a] and on tows (gauge length 115 mm) [RMI 12]. A* is the Anderson–Darling parameter*

Table 7.11 gives as an example a set of failure stresses measured in 4-point bending conditions and the origins of fracture identified with the scanning electron microscope.

σRj (MPa)	Fracture origin
255.78	CHI (S: crack)
259.80	SS (pore)
264.66	S (inclusions)
274.35	S (edge)
275.79	S (agglomerate)
282.00	CHI
292.07	S (pore)
295.09	SS (pore)
299.73	SS (pore)
305.24	S
305.37	S (pore)
305.83	SS (pore)
305.88	SS (pore)
307.31	CHI
308.02	SS (pore)
332.25	V (pore)
311.33	S (inclusion)
312.75	SS (pore)
313.54	S (pore)
314.05	S
323.23	SS (pore)
325.66	SS (pore)
328.49	S (pore)
368.71	V (pore)
333.27	S (pore)

357.88	V (inclusion)
339.41	S (pore)
357.58	V (pore)
351.96	V (pore)
342.62	SS (pore)
342.88	SS (edge: pore)
361.55	V (pore)
346.88	SS (pore) + S
364.30	SS (cracks) + V (pores)
348.33	SS (pore)
348.75	SS (pore)
349.11	S (pore)
364.80	V (pore)
392.77	V (pore) + S
350.50	S (pore)
352.21	CHI
353.97	S (edge: pore)
381.59	V (pore)
355.13	SS (pore)
355.43	SS (pore)
355.43	SS (pore)
358.22	S (pore)
362.40	V (pore) + SS (inclusions)
359.93	S (edge: pore) + SS
372.17	V (pore)
383.46	V (pore)
367.89	SS (pore)
369.73	S (pore)
371.14	SS (pore)
371.47	S
371.63	SS (pore)

371.64	SS (edge: pore) + CHI
372.65	S (pore)
386.58	V (inclusion)
380.05	SS (pore)
384.27	CHI (surface: pore)
385.46	SS (fissure + inclusions) + S
388.64	SS (pore)
409.96	V (pore)
392.48	S (pore)
393.40	SS (pore)
394.30	SS (inclusion)
401.98	SS (pore) + SS (edge: pore)
428.55	S (pore) + SS (pore)

CHI : intersection of botton face and corner; SS: subsurface; S: surface; V: volume

Table 7.11. *Failure stresses measured using 4-point bending tests on silicon carbide. and fracture origins identified using scanning electron microscope*

8

Computation of Failure Probability: Application to Component Design

8.1. Introduction

The statistical-probabilistic models of brittle failure find an important application to the design of components made up of brittle materials, as they allow predictive calculations of failure. They provide a decisive improvement over classical structure calculations by stress analysis. Failure probability is a quantitative and objective criterion of fracture, as well as a coefficient of confidence that can be associated with stresses. It is also an estimate of reliability.

Calculations of failure probability require on the one hand the values of stresses operating on the structure, and on the other hand the statistical parameters characterizing flaw strength distributions pertinent to material. Closed-form expressions of stress states are available only for a few simple loading cases. The calculation of failure probability becomes a complex exercise when the statistical-probabilistic model is based on the distribution function of flaw strengths, when the stress state is multiaxial or transient (thermal shocks), when component geometry is complicated, when several populations of defects coexist within the material, or change under the influence of environment. Computer programs based on the finite element method can calculate the stresses, and can allow difficulties inferred by component geometry and loading conditions to be overcome. A dedicated computer program is then necessary to calculate failure probability.

A number of post processors for failure probability computation were developed. They are based on the Weibull model, the Batdorf model or the multiaxial elemental strength model. This chapter focuses on the CERAM computer program which includes the 2-parameter Weibull equations and the more fundamental multiaxial elemental strength model. The primary function of the code is to calculate the fast-fracture failure probability of macroscopically isotropic ceramic components. The performances of CERAM have been evaluated by comparing results of failure probability computations with experimental cumulative distributions failure data measured on diverse ceramic specimens having various shapes and dimensions.

8.2. Computer programs for failure predictions

The development of methodologies and tools for failure predictions and design of components or structures made of ceramic or brittle material was driven logically by advances of research work on statistical-probabilistic approaches to brittle fracture. Considerable progress was made during the last decades, either within the framework of big programs supported by American federal agencies such as the Department of Energy (DOE), the Japanese Ministry of the Technology and the Industry (MITI), etc., or within the framework of individual actions driven by industrial groups and/or research institutions. The latter situation prevailed in Europe, where there was no specific effort at neither the European Commission nor national government levels to support or drive programs for developing design methodologies for ceramic parts or ceramic structural components. Effort was conducted jointly by companies and research laboratories. Therefore, a certain amount of data are proprietary. However, reports on the activity devoted to reliability analysis, to reliability-based design methodologies, to the development of computer codes for fracture predictions and to component design are available.

Table 8.1 lists the main computer programs for prediction fast fracture failure probability of ceramic parts developed in the world. These computer programs include various statistical theories of fracture and use finite-element analysis output from computer programs in the public domain (2D NIKE, NIKE 3D, DYNA, etc.) or from commercial ones (ANSYS, ABAQUUS and MSC/NASTRAN). Detail can be found in the bibliographical references.

Computer program	Statistical theory	Origin
Japan		TEPCO (Tokyo Electric Power
GFICES		Company)
CCPRO		IHI (Ishikawa-Harima Heavy
		Industries)
		MITI
Europe		
CERAM [LAM 89]	MESM*, Weibull	Battelle Genève
STAU [HEG 91]	Weibull	University of Karlsruhe
CERITS-L [STÜ 91]	Weibull	
CERTUB [SCH 96]	Weibull	
Daimler–Benz [HEM 86]	Weibull	
BRITPOST [SMA 90]	Weibull	
FAILUR [DOR 92]	Weibull	
FEM-LIFTAP [KUS 91]	Weibull	
USA		
CARES [NEM 90]	Batdorf, Weibull	NASA
WESTAC	Weibull	Garret Turbine Engine Company
CERAMIC/ERICA	MESM*	Honeywell Engines, Systems &
[SCH 99]	Weibull	Services
Allison/Cares [KHA 93]	Weibull	Rolls Royce Corporation
SPS Life [BOR 02]		Sundstrand Power System

MESM*: Multiaxial Elemental Strength Model

Table 8.1. *List of computer programs for prediction of fast fracture probability of failure of ceramic components [DUF 03]*

At this stage, developments of computer programs for failure predictions consist on the one hand of translating the models in programming language and on the other hand in introducing diverse functions or modules for the analysis of output results and component design. The methodology covers a number of tasks such as the determination of allowable material properties, the calculation of confidence intervals for the predictions of failure, etc. The developments also involve works of validation and evaluation of the quality of failure predictions using sets of experimental results. In some cases, laws of delayed failure by slow crack growth are introduced for the computation of time-dependent failure probability of thermomechanically loaded components or pieces. More detail can be found in the cited literature.

8.3. The CERAM computer program

A big research program that was aimed at developing a design methodology for ceramic parts was conducted in 1987–1990 by Battelle Europe Research Center (Geneva, Switzerland), with the support of eight large European and Japanese industrial companies involved in automotive industry, aeronautics, energy and engineering. The CERAM software package for reliability analysis of structural parts made up of brittle materials was developed, validated using experimental data (as indicated in section 8.3) and provided to the companies for their own use. CERAM forms part of a specific design methodology for developing reliable ceramic design by computing the failure probability of ceramic components.

The CERAM code includes the multiaxial elemental strength model (see Chapter 4). The Weibull model has been incorporated for comparison purposes, although the Weibull theory of fracture statistics has been found inadequate by several authors in critical applications of structural ceramics such as those involving multiaxial stress states. The Barnett–Freudenthal approximation (principle of independent action of stresses) has been introduced for handling multiaxial fracture (see Chapter 2). In this approach, the principal stresses are assumed to act independently in each principal direction. This approach has been criticized by several authors as it ignores interaction of principal stresses.

CERAM uses two input data files (Figure 8.1):

– stresses at Gauss points of every element of the mesh. This file is generated by a finite element analysis code;

– the statistical parameters of the populations of fracture-inducing defects inherent to the material. Up to seven distinct flaw populations including surface-located or internal flaws can be specified. Each flaw population is characterized by a couple of parameters: the Weibull parameters m and σ_{0w} and the multiaxial elemental strength model parameters m and S_0. The estimation of statistical parameters is discussed in Chapter 7.

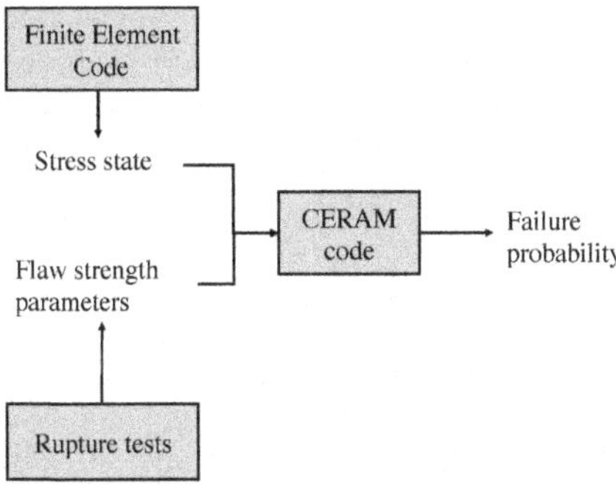

Figure 8.1. *Block diagram for computation of failure probability by means of the computer code CERAM*

The CERAM software package can consider any loading conditions and geometries that can be analyzed by a finite element method code first. Computations can be performed at successive time steps when time-dependent failure probabilities are appropriate as in the case of thermal shocks, cyclic fatigue or dynamic loading. Statistical parameters may depend on time when flaw populations are affected by environment or loading mode. In this case, a time-dependent law must be specified for the statistical parameters. This law can be determined using the methods discussed in Chapter 7, on several sets of failure data obtained after exposure of batches of specimens to environmental or loading conditions during various times. This is an alternative to the approach based on subcritical crack growth that has been introduced in a few computer codes.

CERAM outputs consist of values of failure probabilities of every element of the mesh, the value of component failure probability and maps of distribution of failure probabilities. The value of component failure probability is derived from the product of survival probabilities of mesh elements.

Figures 8.2–8.5 show the examples of failure probability maps obtained for flexural beams, a Brazilian disk, a C-ring and components. The mesh elements with the highest failure probabilities are identified by the red color. They show the critical areas from which fracture should occur preponderantly. The critical area in flexural specimens is the tensile outer surface and the adjacent subsurface. That of Brazilian disk is located at the center subjected to tensile stresses (Figure 8.4). The critical area in components of Figure 8.5 covers a broad region including the inner surface.

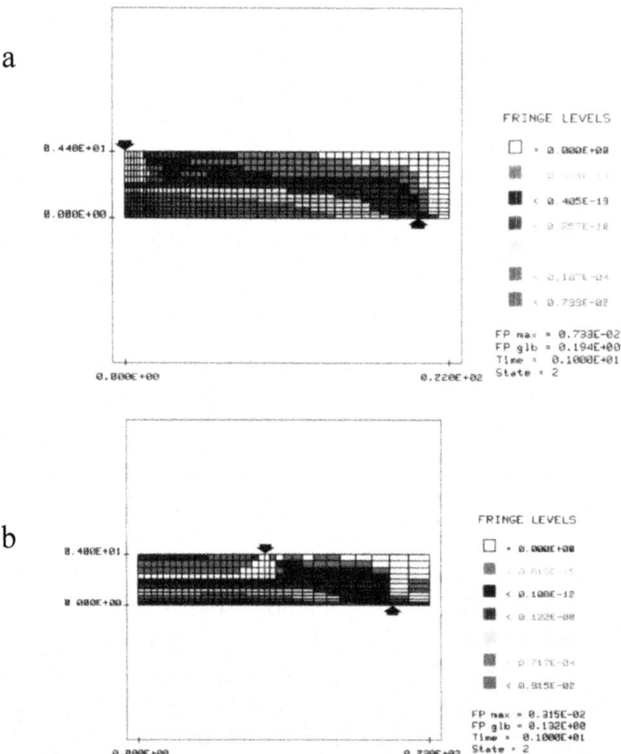

Figure 8.2. *Distribution of fracture probabilities in a 3-point bending specimen of silicon carbide a) and in a 4-point bending specimen of silicon nitride b) computed using the computer code CERAM. Also indicated are the maximal element fracture probability, the specimen failure probability, time and the loading step. For a color version of the figure, see www.iste.co.uk/lamon/ brittle.zip*

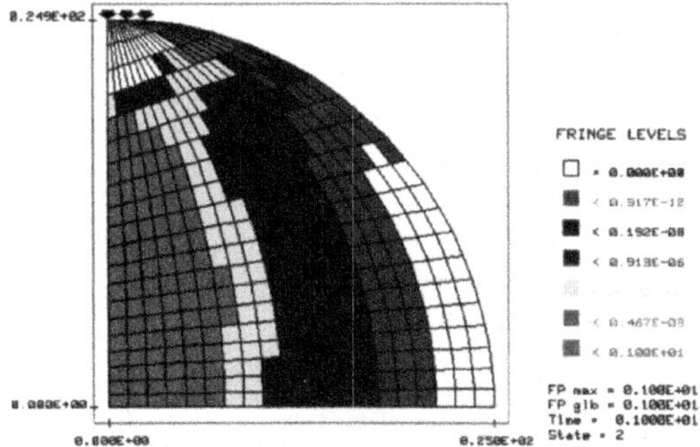

Figure 8.3. *Distribution of fracture probabilities in a disc under diametral compression (Brazilian disk) computed by the CERAM computer code. A quarter of the disk was analyzed for symmetry reasons. For a color version of the figure, see www.iste.co.uk/lamon/brittle.zip*

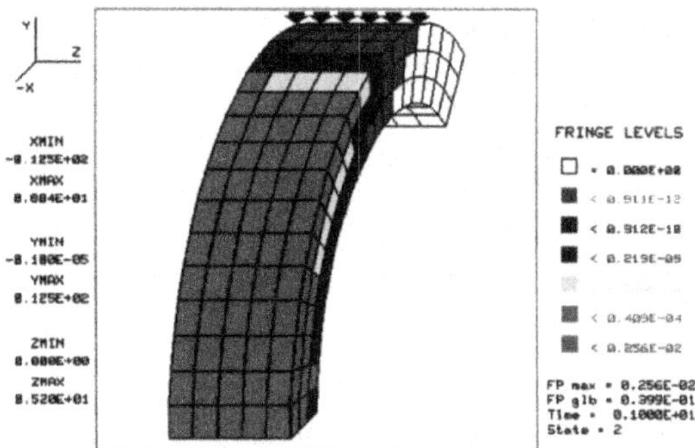

Figure 8.4. *Distribution of fracture probabilities in a C-ring of silicon nitride under diametral compression computed by the CERAM computer code. Half C-ring was analyzed for symmetry reasons. For a color version of the figure, see www.iste.co.uk/lamon/brittle.zip*

The value of component failure probability informs us of the behavior of component. In most previous examples, failure probability (noted by *FP* in the Figures) is between 0 and 1, which means that some specimens will fail whereas the other will survive. The respective proportions of failing and surviving specimens are derived from the value of failure probability. The failure probability of 1 was obtained for the Brazilian disk, which means that all the disks will fail under the present applied load.

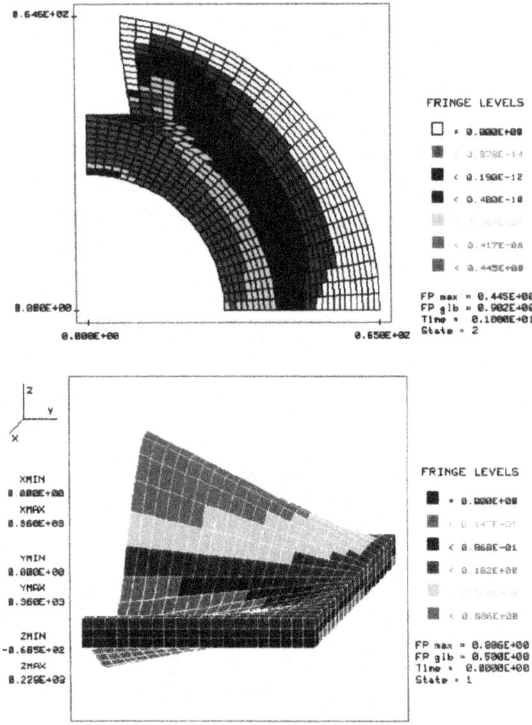

Figure 8.5. *Distribution of fracture probabilities in components computed by the CERAM computer code, taking into account the presence of surface-located and volume-located flaw populations. For a color version of the figure, see www.iste.co.uk/lamon/brittle.zip*

8.4. Validation of the CERAM computer code

The CERAM computer program was validated on a few commercial ceramics by scaling the strength distributions to different specimen sizes and loading modes. The strength distributions were computed using the flaw strength statistical parameters that were estimated on reference specimens. The results were then compared to empirical distributions. Several sets of experimental strength data were measured on various ceramics including silicon nitride, silicon carbide, alumina, possessing single or multiple flaw populations, and under various loading conditions [LAM 88a, LAM 89, LAM 90, LAM 91, LAM 94]:

– 3-point bending with variable span [AMA 88, LAM 88a, LAM 90];

– 4-point bending [AMA 88];

– traction [LAM 89];

– biaxial flexure of disks [LAM 88a];

– compression of C-rings [LAM 90];

– thermal shocks [LAM 90];

– thermal shock heat-up [LAM 91];

– others [LAM 94].

In [LAM 89], 715 specimens of silicon nitride and silicon carbide were prepared for seven different test configurations. Some of the case studies are discussed in Chapter 9.

The influence of the following factors was examined [LAM 94]:

– the probability estimator for determination of empirical distribution;

– uncertainty and variability of statistical parameters;

– the probabilistic model;

– the finite element mesh;

– the accuracy of stress analysis;

– the definition of failure stress in the presence of stress gradient.

It was shown that with reproducible ceramics and appropriate estimates of flaw strength parameters, the failure probabilities predicted using CERAM were in good agreement with experimental failure data. The multiaxial elemental strength model gave more accurate predictions than the Weibull model did. Typically, the

Weibull model underestimated failure probabilities and, in some loading cases involving shearing stresses, led to significant discrepancies.

The estimation of statistical parameters is a critical step. Failure predictions are highly sensitive to estimates of statistical parameters, due to the power form of underlying flaw strength density function. The influence of failure criterion is of secondary importance when comparing with that of uncertainty in statistical parameters. Thus, calculations have shown that 10% uncertainty in scale factor is responsible for 15–100% discrepancy in failure probability predictions.

The CERAM-based method of estimation of statistical parameters (see Chapter 7) appeared to be superior to the conventional techniques (linear regression, maximum likelihood and mean strength) when comparing failure predictions with the different statistical parameters. This is due to the performance of numerical analysis for the determination of stresses, and then failure probabilities. Accuracy in stress analysis requires refinement of mesh in those parts subjected to stress gradients, and optimization according to the rule that prevails for sound finite element analysis.

The Weibull scale factor was found to be greater than the multiaxial elemental strength model one, as a result of values of I_S (…) or I_V (…) < 1: $S_0 / \sigma_0 > 0.9$. Therefore, we cannot substitute one scale factor for another.

8.5. CERAM-based ceramic design

Designing methods based on failure probability computations is more efficient than stress analyses, as values of failure probability provide criteria for both failure and reliability of component in service. Stress analysis-based methods mimicking metallic approach are not sufficient for brittle materials, since the computed stresses cannot be compared as such to an intrinsic material fracture strength. As discussed in this book, the fracture strength cannot be characterized by a single value. Instead, it depends on several parameters, including the loading mode and the size of specimen. The computed stresses operate on small elements that are several orders of magnitude smaller than the test specimens used for measuring material strength. The stress analysis is not capable of determining whether the material is suitable. It provides only qualitative information like the location of high stresses. Due to the lack of predictive analysis, design with ceramics has been empirical. This led to many component failures in the past. The design of component evolved from these failures. It consisted of first replacing metal components peacemeal rather than adapting the system according to ceramic features. Figure 8.6 shows the different steps of metallic design method: given the shape and dimensions of metallic component, manufacture and testing of component. If the component fails during tests, stress analysis and fractographic inspection are carried out to remedy the failure. The modifications to be brought to the shape of part are decided according to the distribution of stresses. The decrease of stress concentrations is sought. A new component with improved geometry is then made and tested, and the behavior is analyzed as in the first step. The iterative procedure needs to be carried out until the failure is overcome. This approach is costly and time-consuming. The final stage may never be reached when it comes to material selection.

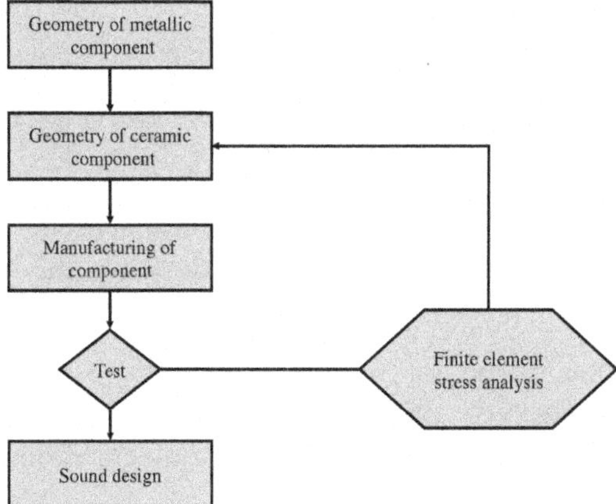

Figure 8.6. *Metallic approach to designing ceramic components*

Figure 8.7 shows the alternative design method which is based on calculation of failure probabilities using a computer software such as CERAM. The following preponderant factors are examined successively:

– the conditions of service;

– the geometry of part;

– and the properties of material which are specified in two files: thermoelastic properties for the finite element analysis of stresses and statistical parameters of flaw populations for the calculation of failure probability.

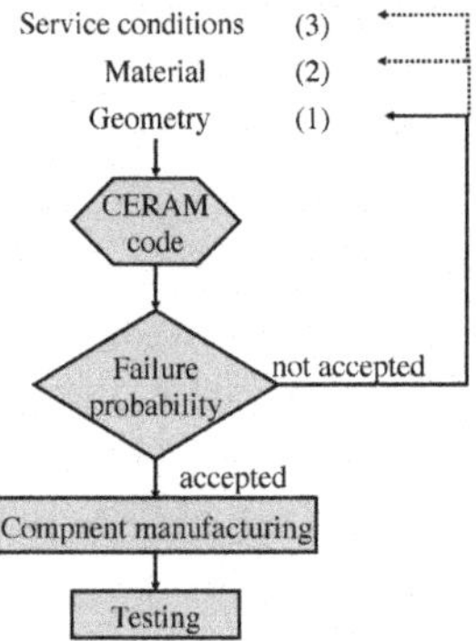

Figure 8.7. *Design methodology of ceramic components based on failure probability computations using the CERAM computer code*

The outputs of failure probability computations allows for the diagnosis of design weaknesses, and can guide the decisions to be made:

– either the value of failure probability is sufficiently low according to various requirements (performances). The design is appropriate and the component can be manufactured and tested;

– or the value of failure probability is too high. Then the design of the part must be improved to reduce the failure probability to an allowable level.

In the second alternative, component geometry must be modified in the first place. This operation is guided by the distribution of failure probabilities. Design improvement and associated computations are iterated until allowable component failure probability is obtained. If this latter is not obtained after a reasonable amount of runs, it is then advisable to modify one of the two other dominating factors. Alternative material properties can be selected. If the value of failure probability obtained for the new material is acceptable, the problem is solved. Should the opposite occur, the computations should be run for a different material. The inverse method is to fix that allowable failure probability and to identify the corresponding properties. A method is presented in the following section.

The computations can yield the conclusion that no suitable ceramic is available. Thus, it remains to modify the conditions of service and to follow the same way for the new conditions of service. If the exploration is in vain, the conclusion can be drawn that ceramic materials cannot suit for the envisaged application. Composite materials may be an interesting alternative.

8.6. Relation test specimen/component: identification of allowable material properties

The equations of failure probability allow the failure stresses measured on test specimens to be related with the allowable service stress and failure probability of component. These relations can be obtained using the approach presented in Chapter 6 on size and stress state effects on strength and summarized by charts of practical use.

According to equation [7.1], component failure probability is expressed as:

$$P_p = 1 - \exp\left[-\frac{V_p}{V_o} K_p \left(\frac{\sigma_p}{\lambda_o}\right)^m\right] \qquad [8.1]$$

where subscript p refers to “part”.

A similar equation gives specimen failure probability (with subscript ep for “specimen”):

$$P_{ep} = 1 - \exp\left[-\frac{V_{ep}}{V_o} K_{ep} \left(\frac{\sigma_{ep}}{\lambda_o}\right)^m\right] \qquad [8.2]$$

Let us recall that the constants K_p and K_{ep} depend on stress state and statistical-probabilistic model. Both K_p and K_{ep} are calculated by means of CERAM. σ_p represents the maximum stress in the part. σ_{ep} is the maximum stress in the test specimen. From the ratio of logarithms of expressions [8.1] and [8.2], it comes the following relation between σ_p and σ_{ep}:

$$\sigma_{ep} = \sigma_p \left[\frac{V_p}{V_{ep}} \cdot \frac{K_p}{K_{ep}} \cdot \frac{L_n(1-P_{ep})}{L_n(1-P_p)}\right]^{1/m} \qquad [8.3]$$

From equation [8.3], the following relation is then derived between the average failure stress measured on a lot of test specimens and the component allowable stress for failure probability P_p. To simplify the presentation, it is assumed that the average is obtained for a probability of break equal to $P_{ep} = 0.5$ (which is not strictly exact).

$$\bar{\sigma}_{ep} = \sigma_p \left[\frac{V_p}{V_{ep}} \cdot \frac{K_p}{K_{ep}} \cdot \frac{L_n(0{,}5)}{L_n(1-P_p)}\right]^{1/m} \qquad [8.4]$$

The corresponding chart (Figure 8.8) can be read in two different manners:

– when component requirements (stress on component σ_p and associated probability P_p) are available, the chart yields the strength of specimen made of same material as component. Thus, the suitable material for a safe use of component can be identified. On the example of Figure 8.8, for stress on component σ_{p1}, the strength of appropriate material must be larger than σ_{ep3} when component failure probability is P_{p3}. If higher probabilities are targeted, larger material strengths are necessary such as: $\sigma_{ep1} > \sigma_{ep2} > \sigma_{ep3}$ depending on component probability $P_{p1} > P_{p2} > P_{p3}$. The safe selection would be material having strength $\geq \sigma_{ep1}$;

– conversely, material strength σ_{ep} is known, the values of allowable component failure stresses and probability can be predicted.

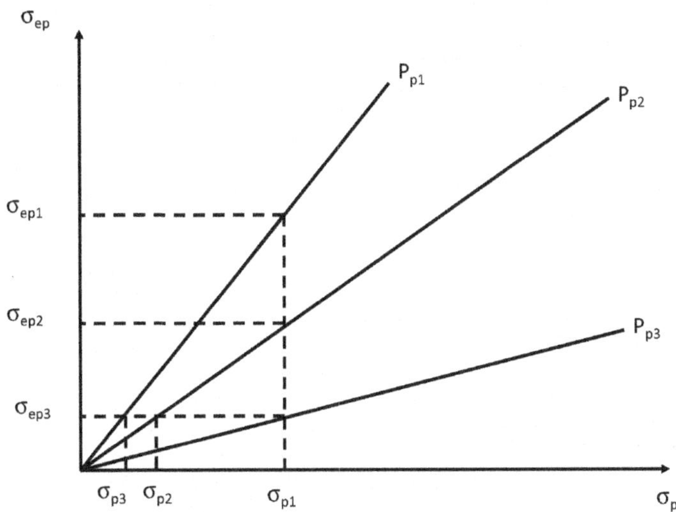

Figure 8.8. *Mean failure stress of test specimens ($\bar{\sigma}_{ep}$) versus allowable component stress σ_p, for various values of component failure probability P_p: $P_{p1} > P_{p2} > P_{p3}$ (logarithmic coordinates)*

A chart as shown in Figure 8.8 must be established for each considered part, test specimen geometry and loading mode using equation [8.4]. Equation [8.4] can also be used to optimize stress on component for a given material. For this purpose, it is advisable to look for the minimum of the function $[K_p]^{1/m}$, so as to increase σ_p at a given value of $\bar{\sigma}_{ep}$. Thus, m has to be as large as possible.

Such a chart appears as a powerful tool to correlate directly the resistance of the material measured on test specimens, and the conditions of service and the reliability of parts. A similar approach represented in Figure 8.9 was adopted for the development of SiC turbines blades by the companies Carborundum (producer) and Allison Gas Turbine (user) in the 1980s [LIN 83]. The chart of Figure 8.9 is a particular case with regard to Figure 8.8, because the probability of survival and the conditions of service of turbines are specified. σ_p and P_p being fixed, equation [8.4] is reduced to the following expression:

$$\bar{\sigma}_{ep} = [K(P_p)]^{1/m} \quad [8.5]$$

with

$$K(P_p) = \sigma_p \left[\frac{V_p}{V_{ep}} \cdot \frac{K_p}{K_{ep}} \cdot \frac{Ln(1-0,5)}{Ln(1-P_p)} \right]^{1/m} = \text{constant} \quad [8.6]$$

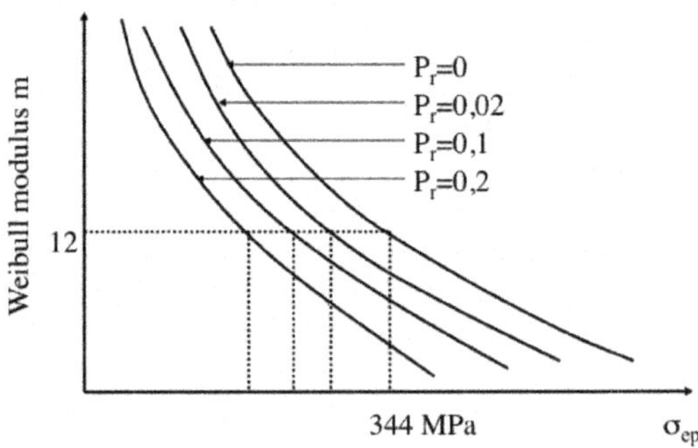

Figure 8.9. *Weibull modulus versus 4-point flexural strength of silicon carbide for various values of rejection rate for turbine blades under critical starting up conditions*

The chart was constructed to estimate the characteristics of SiC so that the turbine blades can withstand the critical loading conditions at the starting up of the engine when rotation speed reaches 83% of the cruising speed and temperature, respectively, 83 and 78% of those at cruising speed. The survival probability of component under these conditions was set to P_S = 0.976. The stresses generated by these conditions were calculated by the finite element method and the factor K_p of equation [8.6] was determined using the Weibull model. The average values of the failure stress of test specimens were calculated for various values of m and of the rate of rejection P_r, defined as $P_p = P_r + (1 - P_s)$ where P_s is the probability of survival specified above (Figure 8.9). Reference test specimens and corresponding lots of turbine blades were produced at the same time. From the values of failure stresses of specimens, turbine blade performances can be evaluated on the chart (Figure 8.9). It can be noted that for m = 12, test specimen failure stress superior to 344 MPa is required so that $P_r = 0$. The rejection rate increases with weaker specimen failure stress so that design of a proof test may be necessary for test specimen failure stresses close to 344 MPa.

8.7. Determination of statistical parameters using CERAM

The CERAM computer program also enables us to determine the scale factor λ_0, which is known to influence strongly the quality of failure predictions. The principle of the method is simple. The failure probability of component or test specimen is computed, for a value of shape parameter m estimated independently, and then it is compared to available empirical value. An arbitrary value of λ_0 is used for the calculations, the scale factor is derived using equation [7.15], the demonstration of which is immediate. Indeed, taking logarithms of equation [7.1], it comes for failure probability:

$$-\mathrm{LnLn}\,\frac{1}{1-P_{CERAM}} = -\,\mathrm{m}\ \mathrm{Ln}\,\lambda_{0A} + \mathrm{Ln}\,[\,\frac{V}{V_O}K\sigma_{ref}^{m}\,] \qquad [8.7]$$

where P_{CERAM} is the failure probability computed using CERAM, for the value λ_{0A} of scale factor. The corresponding equation of empirical failure probability is:

$$-\mathrm{LnLn}\,\frac{1}{1-P_{\exp}} = -\,\mathrm{m}\ \mathrm{Ln}\,\lambda_{o} + \mathrm{Ln}\,[\,\frac{V}{V_O}K\sigma_{ref}^{m}\,] \qquad [8.8]$$

where λ_o is the unknown value of scale factor. Equating expressions [8.7] and [8.8] yields [7.15].

This method of determination of statistical parameters has several advantages. First, it can be integrated into design methodology. Then, it can be employed with failure data generated on test specimens of complicated geometry, as well as with parts when the reproducibility of material and statistical parameters during scaling up is an issue. Finally, it is recommended when models such as the multiaxial elemental strength model are used.

8.8. Application to multimaterials and composite materials

In the case of assemblies of ceramic or fragile materials, the determination of statistical parameters does not raise practical problem when it is possible to prepare test specimens of every material having such dimensions that they can be tested. Then, the statistical parameters estimated for the various constituents have to be specified in the appropriate files of input data (Figure 8.1). In the case of ceramic/metal assemblies, the calculation of failure probability is pertinent only for the ceramic constituent, unless the metallic part is brittle.

The contribution of joining materials (ceramic adhesives) may have to be taken into account. To our knowledge, this problem has hardly been treated. However, it seems that it can be handled. The statistical parameters of fracture-inducing flaw populations have to be estimated. Then, as previously, it is necessary to check whether the concept of weakest link hypothesis is not violated, and whether joint fracture is actually brittle. In the case when joint fracture is not brittle, the failure probability will characterize the initiation of a crack.

The ceramic composites made of ceramic strengthened by ceramic fibers are regarded as structures. They combine several of the problems evoked above. They are made of an assembly of materials, which fracture at local strength scale: microcracks are created which do not cause the failure of composite material. The weakest link concept is thus pertinent only on a local length scale, but not on the length scale of the structure. The probabilistic models of brittle fracture thus apply to microcrack creation from heterogeneities having random distribution. This will be discussed in Chapter 10.

8.8.1. *Prediction of damage by microcracking in ceramic composites*

Probabilistic models of microcracking of matrix in ceramic composites are discussed in Chapter 10. Here, it is only a question of showing how the probability of creation of microcracks can be computed with the CERAM computer code.

The matrix and the fibers are the elemental constituents of composite. They are generally isotropic materials. Their properties as well as the statistical parameters of flaw strength distributions are required for the computations of constituent failure probabilities. Figure 8.10 shows an example of finite element mesh for a cell of 2D woven SiC/SiC composite made by chemical vapor infiltration (CVI). The cell reproduces the composite microstructure made up of woven infiltrated tows, large pores at tow crossing and a uniform layer of matrix on the fiber preform (referred to as interply matrix). Extensive inspection of test specimens under tensile load has shown that microcracking proceeds progressively and successively in the interply matrix, in the matrix of transverse tows (perpendicular to loading direction) and finally in the matrix of longitudinal tows (parallel to loading direction). The matrix flaw strength statistical parameters were estimated from the results of tensile tests on microcomposite (matrix reinforced by a single filament) and minicomposite specimens (matrix reinforced by a tow of several hundreds of filaments) (Table 8.2).

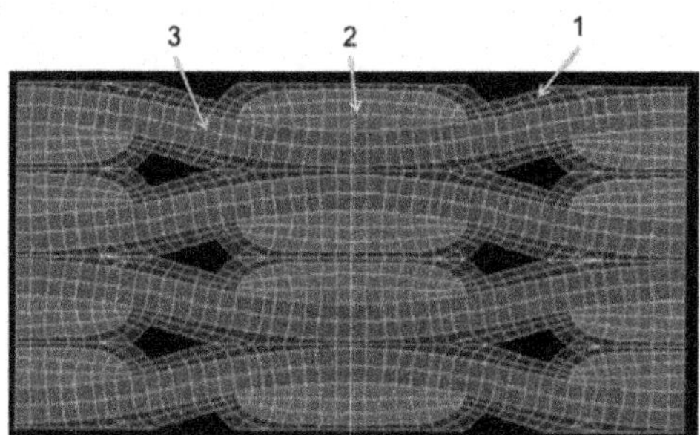

Figure 8.10. *Example of finite element mesh for 2D woven SiC/SiC composite: (1) interply matrix, (2) transverse matrix infiltrated tows, (3) longitudinal matrix infiltrated tows. For a color version of the figure, see www.iste.co.uk/lamon/brittle.zip*

	Module m	S_0 (MPa)
Interply matrix	4.9	1.6
Matrix in transverse tows	4.9	0.7
Matrix in longitudinal tows	6.2	7.3

Table 8.2. *Flaw strength statistical parameters for the CVI SiC matrix*

Figure 8.11 shows the distributions of failure probabilities of mesh elements computed at various load increments. The highest failure probabilities indicate the most likely locations for microcrack initiation. When the applied strain is increased, they are obtained first at macropore tip, then at the tip of microcracks and in the matrix in the transverse and in the longitudinal tows.

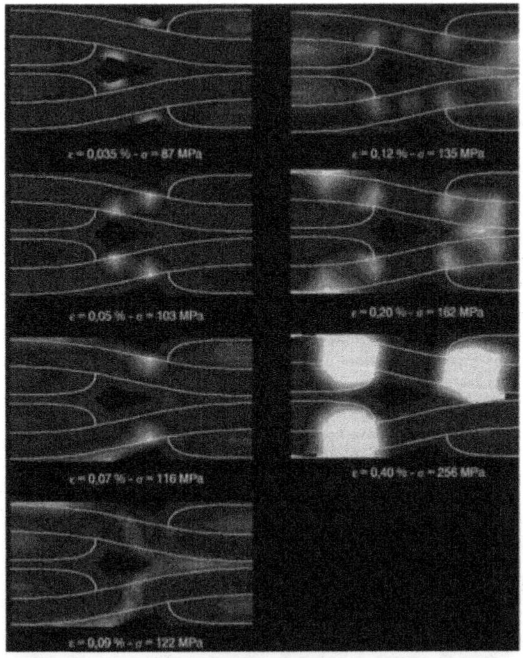

Figure 8.11. *Distribution of failure probabilities in a cell of 2D woven SiC/SiC composite under tensile load parallel to longitudinal fiber axis. For a color version of the figure, see www.iste.co.uk/lamon/brittle.zip*

The maximum value of 1 for element failure probability indicates that microcracking does occur in that element, denoted by j for the sake of clarity. In order to reproduce the microcrack in the mesh, the node of the element j that is common to the two elements with the highest failure probabilities is split into two nodes. After a step of node splitting, the stresses and failure probabilities are computed at constant applied load. If failure probability is again 1 in an element, one more node is split. Computations at constant load after node splitting are iterated as long as failure probability is 1 in an element.

The mesh must be constructed nicely in order to minimize uncertainty in stresses. It must be finer in the zones of expected stress gradient. For the above mentioned 2D SiC/SiC composite, the location of microcracking and the corresponding applied load was predicted satisfactorily [GUI 96, LAM 98]. Experimental results confirmed predictions: difference less than 0.005% was obtained between the applied strains at which the microcracks formed during the tests and the calculated values (Table 8.3). Figure 8.12 shows the predictions of the microcrack network under increasing tensile load. The crack pattern is similar to that observed during experiments. Figure 8.13 shows that the damage behavior of the cell is well described by the Young's modulus degradation curve derived from computations.

Microcraking	Experimental strain (%)	Predicted strain (%)
From macropores	0.035	0.04
In transverse tows	0.09	0.11
In longitudinal tows	0.19	0.2

Table 8.3. *Comparison of predictions and experimental results of matrix damage and corresponding applied strains*

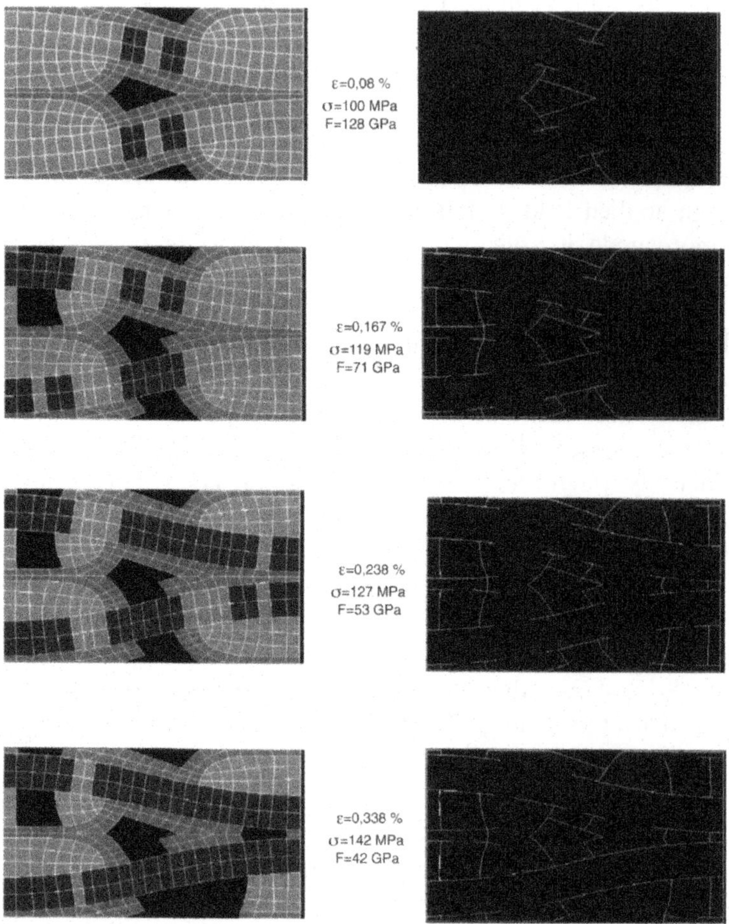

Figure 8.12. *Distributions of failure probabilities and derived networks of micro cracks in a cell of 2D woven SiC/SiC composite under deformation applied parallel to longitudinal fiber axis. Also indicated are values of stresses on cell boundaries and cell Young's modulus. For a color version of the figure, see www.iste.co.uk/lamon/brittle.zip*

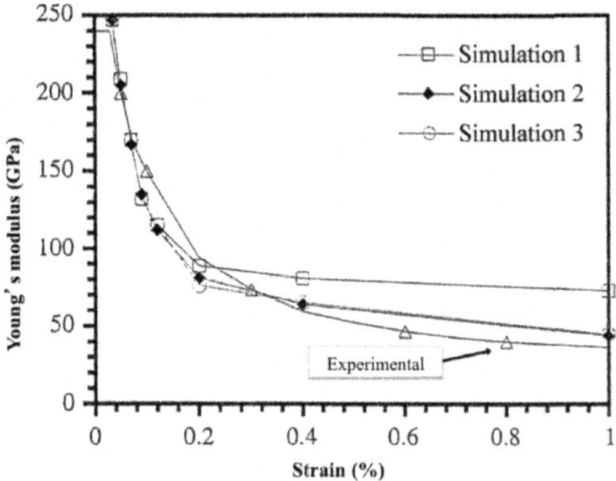

Figure 8.13. *Young's modulus versus applied strains for 2D woven SiC/SiC composite under tensile load parallel to longitudinal fiber axis: comparison of results of experiments and computations for different ply arrangements. simulation 1: shifted plies, no fiber/matrix debonding; simulation 2: shifted plies, with fiber/matrix debonding; simulation 3: parallel tow undulation, with fiber/matrix debonding.*

8.9. Conclusion

The probabilistic post-processor CERAM that uses the physics-based multiaxial elemental strength model is able to predict the failure of structures under complex loading cases. It can take into account the contribution of multimodal populations of defects situated on surface and inside the material, or that are affected by environment. CERAM is also used to determine the statistical parameters, from the results of failure tests on specimens or parts. CERAM computations were evaluated on a number of case studies, which are discussed in the Chapter 9.

CERAM, as well as the probabilistic post-processors, is the heart of a methodology of component design appropriate to brittle materials. This methodology is based on the calculation of failure probability. The geometry of structure, the characteristics of material and the conditions of service can be adapted according to the level of failure probability dictated by diverse performances defined for the structure. The approach is iterative, and it can be inverse.

CERAM can be applied to assemblies, multimaterials and ceramic composites. However, it is important to check that the weakest link concept applies. The fracture of constituents is brittle but local because the microcracks that form in the matrix or that break the small diameter filaments are arrested. As a result, the failure probability characterizes the probability of formation of a microcrack from a defect. This approach thus allows predictions of either the formation of the first crack or the damage by microcracking.

9

Case Studies: Comparison of Failure Predictions Using the Weibull and Multiaxial Elemental Strength Models

9.1. Introduction

In this chapter, the method of calculation of failure probability is applied to various loading cases. Both fundamental steps of statistical parameter estimation and failure probability calculation are conducted by using the finite element analysis of stress state and the post-processor CERAM. Analytical approaches are given for comparison when they are available or possible. Indeed, when the loading mode is different from that of traditional elementary tests such as uniaxial traction or 3-point and 4 point-bending, the development of exact expressions of the stress field is made difficult. Moreover, the physics-based models require the manipulation of complex equations. The analytical equations can thus lead to erroneous statistical parameters and wrong failure predictions.

The case studies address a diversity of loading modes produced in laboratory to determine the statistical parameters, and also to activate the appropriate populations of defects. It is, indeed, necessary to estimate the statistical parameters of all the populations of fracture-inducing defects. The loading modes discussed in this chapter involve some flexion and thermal shocks. The statistical parameters are estimated from empirical distributions of failure strengths measured on tests specimens. They are primary input data for the computation of failure probability under different loading conditions. The case studies illustrate:

– the effects of scale and stress state when failure is caused by a single population of defects located either at the surface, or within the specimen, or by a bimodal population of surface-located and internal defects;

– the failure under transient stress field generated by thermal shocks and thermal fatigue, for unimodal or bimodal flaw populations.

The multiaxial elemental strength model was mainly used since it is at the heart of this book and the post-processor CERAM. The Weibull model was used for comparison purposes. Predictions using the Batdorf model were reported from the literature [SHE 84]. The comparison of analytical and numerical predictions highlights errors resulting from approximations in analytical approaches. Enough detail is given so that the calculations can be reproduced as exercises or so that these examples provide a basis to failure analyses.

9.2. Predictions of failure under flexural load

Flexural testing presents numerous advantages. The implementation of the test is easy, the geometry of test specimens is simple. This test is recommended for determination of strength of brittle materials. It offers a multitude of possibilities: a variety of experimental conditions and specimen geometries can be selected, so that various stress states can be generated.

9.2.1. *Unimodal population of surface defects*

The example discussed in this section was treated analytically by Shetty *et al.* [SHE 84]. The authors produced empirical distributions of failure stresses under diverse conditions of flexion, estimated the Weibull statistical parameters from 4-point bending failure data and calculated the probability of failure for alternative conditions of flexion using the models of Weibull and Batdorf. The modes of flexion generated more or less severe stress gradients (3-point or 4-point bending) and an equibiaxial stress state (biaxial flexure of disks). The tests specimens were made of alumina. The experimental conditions are described in Table 9.1 and Figure 9.1. There are several techniques of biaxial flexion [KIR 67]. Shetty employed the uniform-pressure-on-disk technique. The test consists of loading a disk specimen which is supported along a concentric line support near its periphery, with lateral uniform pressure (Figure 9.1).

Loading mode	Specimens	Dimensions	Test conditions
3-pt bending	beams	2d = 2.5 mm b = 5 mm L = 38 mm	2 l = 32 mm: support span length
4-pt bending	beams	2d = 2.5 mm b = 5 mm L = 38 mm	$2\,l_2$ = 19 mm: loading span length $2\,l_1$ = 32 mm: support span length
Biaxial flexure	discs	e = 2.5 mm $2r_2$ = 31.75 mm	$2r_1$ = 25.4 mm support diameter

2d = beam thickness
e = disc thickness
b = width
L =beam length
$2r_2$ = disc diameter

Table 9.1. *Conditions for the tests on alumina [SHE 83, SHE 84]*

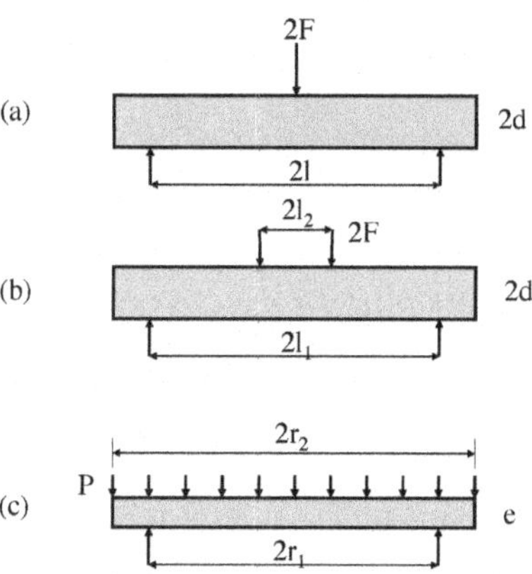

Figure 9.1. *Notations and symbols for the flexural tests: a) 3-point bending, b) 4-point bending, c) biaxial bending of disks*

Shetty's analytical approach is presented at first, followed by the estimation of statistical parameters and failure predictions using the CERAM computer code.

9.2.1.1. *Empirical distributions of failure stresses*

The failure stresses are derived from the force or the pressure at specimen fracture using the following theoretical expressions of reference stresses (maximum stress in the specimen). The probability estimator $P_i = \frac{i-0.5}{n}$ recommended in Chapter 7 was used for the construction of empirical distribution of strengths.

– 4-point bending:

$$\sigma_{ref} = \frac{3F_{max}(l_1 - l_2)}{2bd^2} \qquad [9.1]$$

– 3-point bending:

$$\sigma_{ref} = \frac{3F_{max}l}{2bd^2} \qquad [9.2]$$

– biaxial flexure:

$$\sigma_{ref} = \frac{3P_{max}r_1^2}{8e^2}\left[2(1-\nu)+(1+3\nu)(\frac{r_2}{r_1})^2 - 4(1+\nu)(\frac{r_2}{r_1})Ln(\frac{r_2}{r_1})\right] + \frac{(3+\nu)}{4(1-\nu)}P_{max} \qquad [9.3]$$

where $2F_{max}$ is the maximum force, P_{max} is the maximum pressure and ν is the prison coefficient. The other symbols are defined in Figure 9.1 and Table 9.1.

The specimens did exhibit brittle failure mode, the tests had been conducted in an atmosphere of dry nitrogen so as to avoid subcritical crack growth of defects. The maximal values of forces and pressure during tests are thus appropriate characteristics of specimen brittle fracture. The empirical distributions of failure stresses are shown in Figures 9.2, 9.3 and 9.4, as Weibull plots, that is plots of LnLn[1/(1-P)] versus $Ln\sigma_{ref}$. They fit satisfactorily single lines that are quite parallel, which suggests that a single population of defects was responsible for fracture. The fractographic examination of the test specimens showed that fracture originated from surface located flaws.

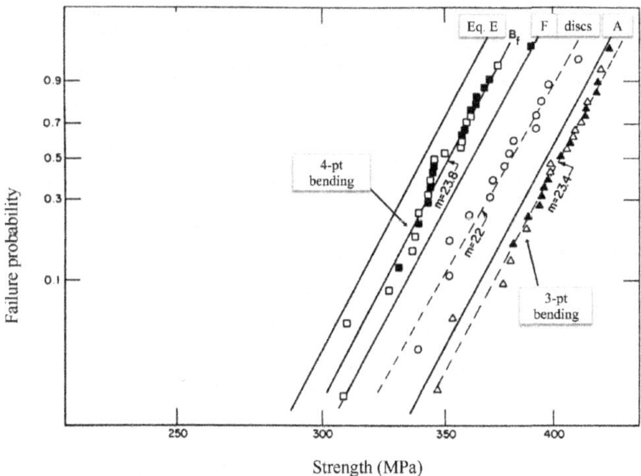

Figure 9.2. *Predictions of failure stresses using the Weibull model for the 3-point bending bars (equation (A), Table 9.2) and for the disks (equations (E) and (F), Table 9.2) for the statistical parameters estimated from the Weibull plot of 4-point bending fracture stresses [SHE 84]. Solid line B_f represents experimental base line for predictions*

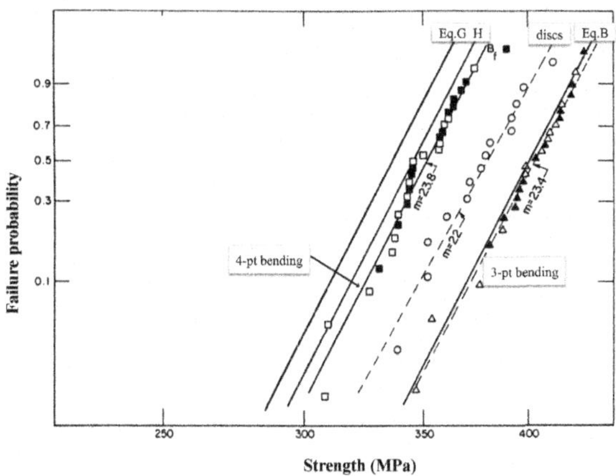

Figure 9.3. *Predictions of failure stresses using the Batdorf model for the 3-point bending bars (equation (B), Table 9.2) and for the disks (equations (G) and (H), Table 9.2) for the statistical parameters estimated from the Weibull plot of 4-point bending fracture stresses [SHE 83]. Solid line B_f represents experimental base line for predictions*

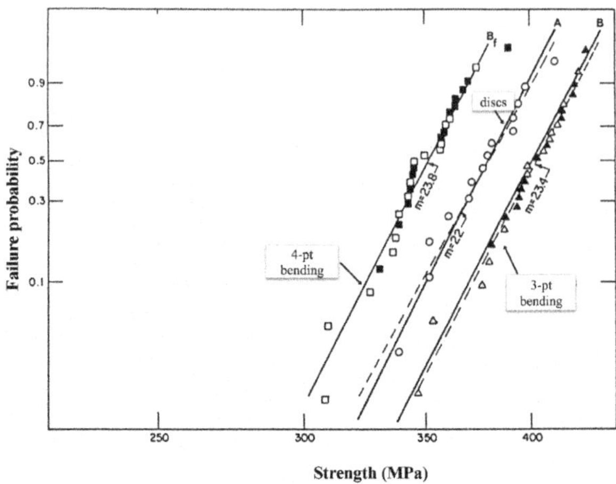

Figure 9.4. *Predictions of failure stresses computed using the CERAM computer code (Multiaxial elemental strength model) for the 3-point bending bars and the disks, for the statistical parameters estimated from the Weibull plot of 4-point bending fracture stresses established by Shetty [SHE 84]. Solid line B_f represents experimental base line for predictions*

9.2.1.2. *Analytical calculation of failure probability [SHE 83, SHE 84]*

According to the surface-location of fracture-inducing flaws, probability of fracture from the lateral and bottom faces of bars was calculated. The expressions of stress state in the 3-point and 4-point bending bars are given in Chapter 5.

The stress state in the tensile bottom face of disks consists of radial and tangential components that are maximum and equal at the disk center and decrease along a radius according to the following relations:

$$\sigma_r = \sigma_{max} \; [1-\alpha(\frac{r}{r_1})^2] \qquad [9.4]$$

$$\sigma_t = \sigma_{max} \; [1-\beta(\frac{r}{r_1})^2] \qquad [9.5]$$

where σ_{max} is given by equation [9.3] ; parameters $\alpha = \frac{3P(1+\nu)r_1^2}{8\sigma_{max}e^2}$ and $\beta = \frac{3P(1+3\nu)r_1^2}{8\sigma_{max}e^2}$ define the stress gradient along a radius.

The expressions of failure probability are derived from general equations given in Chapters 2 and 4, with the symbols referring to volume replaced by those referring to surface: equations [2.17] for uniaxial stress state and equations [2.22] and [2.29] for multiaxial stress state for the model of Weibull; equation [4.22] for the model of Batdorf (Table 9.2).

	Weibull model	Batdorf model
3-pt bending	$\frac{2bl(m+1)+4ld}{(m+1)^2}\left(\frac{\sigma_{max}}{\sigma_o}\right)^m$ (A)	$\frac{\dot{k}\,\sigma_{max}^m}{\lambda m+1}\left\{\frac{2bl(m+1)+4ld}{(m+1)^2}\right\}$ (B)
4-pt bending	$\frac{(m\frac{l_2}{l_1}+1)\,(2bl_1(m+1)+4l_1d)}{(m+1)^2}\left(\frac{\sigma_{max}}{\sigma_o}\right)^m$ (C)	$\frac{\dot{k}\,\sigma_{max}^m}{\lambda m+1}\left\{\frac{(m\frac{l_2}{l_1}+1)(2bl_1(m+1)+4l_1d)}{(m+1)^2}\right\}$ (D)
Biaxial flexure	$\left(\pi r_1^2\right)\left(\frac{\sigma_{max}}{\sigma_o}\right)^m Lu_1$ (E) $\left(\pi r_1^2\right)\left(\frac{\sigma_{max}}{\sigma_o}\right)^m \frac{1}{m+1}\frac{\alpha+\beta}{\alpha\beta}$ (F)	$\frac{\dot{k}\,\sigma_{max}}{2m+1}\left\{\frac{\sqrt{\pi}m\Gamma(m/2)\pi r_1^2}{(m+1)\Gamma(m+1/2)\alpha}\right\}$ (G) $\frac{\dot{k}\,\sigma_{max}}{m+1}\left\{\frac{\sqrt{\pi}\Gamma(m+2)\pi r_1^2}{2(m+1)\Gamma[(m+1)/2]\alpha}\right\}$ (H)

Table 9.2. *Expressions of risk of rupture B (P=1 – exp – B) established by Shetty et al. [SHE 83, SHE 84]. Equation (E) was obtained using the multiaxial Weibull's formulation. Expression of loading factor Lu_1 in equation (E) is detailed in [SHE 83]. The expression (F) was obtained for the principle of independent action of stresses. Equation (G) was obtained for the normal stress criterion, equation (H) for the strain-energy-release rate criterion that assumes coplanar crack extension. In equations (B) and (D), λ = 1 for the normal stress criterion, λ = 2 for the strain-energy-release rate criterion that assumes coplanar crack extension. Equations (G) and (H) provide lower bounds. Upper bounds are obtained by substituting β for α in (G) and (H)*

Weibull's theory implicitly assumes a normal stress fracture criterion, which is inadequate to treat fracture from cracks oriented at arbitrary angles to principal stresses. For the analysis of disks using the model of Batdorf, two fracture criteria were employed: on one hand the most recognized and the simplest criterion for brittle materials, based on the normal stress σ_n:

$$\sigma_e = \sigma_n \geq \sigma_{cR} \qquad [9.6]$$

on the other hand a strain-energy-release rate criterion that assumes coplanar crack extension:

$$\frac{\sigma_n^2}{\sigma_{cR}^2} + \frac{\tau^2}{\sigma_{cR}^2} = 1 \qquad [9.7]$$

where σ_n is the normal stress component, τ is the shear stress parallel to a crack and σ_{cR} is a critical stress defined in Chapter 4.

The Weibull modulus was estimated from Weibull plots using linear regression method (Chapter 7). The scale factor has been estimated from the strength distribution derived from 4-point bending results (Table 9.3). It can be noted in Table 9.3 that close values of the Weibull modulus have been obtained, which confirms the presence of a single population of fracture-inducing flaws as identified by fractography. The values of the Weibull modulus are rather high for a ceramic. Shetty indicated that special care was taken in the qualification of the strength tests to eliminate sources of strength variations other than those arising from the variability of strength-controlling flaws. A larger scale factor was determined for the Batdorf model. Figures 9.2 and 9.3 show the failure probability distributions computed for the 3-point bending bars and for the disks using the statistical parameters estimated from 4-point bending strengths. Strengths of disks were underestimated significantly. The discrepancy is reduced with the criterion defined by equation [9.7].

9.2.1.3. *Numerical computation of failure probability*

– *Finite element analysis of stresses*

For symmetry reasons, planar analysis in vertical midplanes of half specimens was performed. A mesh of 400 quadrangular elements was constructed. The following elastic properties of alumina were used: Young's modulus E = 316 GPa, Poisson coefficient $\nu = 0.22$.

– *Estimation of statistical parameters*

The scale factor of each strength distribution has been estimated using the method discussed in Chapter 7, section 7.2.4. Probability of surface-originated fracture has been computed using the CERAM computer code for the shape parameters given in Table 9.3, for an arbitrary value of scale factor (taken to be the Weibull one reported in Table 9.3), and for a particular value of force on bending bar, and pressure on disk (Table 9.4). The corresponding values of σ_{ref} that have been computed using the finite element method are given in Table 9.4. They are related to empirical values of failure probability P_{exp} through the empirical distribution of strengths. The values of scale factors have then been determined using equation [7.15]. From the results given in Table 9.3, it can be noted the following interesting features for the values of scale factor of the multiaxial strength model:

– first, they show quite no dependence on loading mode as expected since they are invariant in the presence of a single population of pre-existing flaws in the material. By contrast, those of Weibull depend on stress state;

– second, they are smaller than the Weibull scale factors, as pointed out in Chapter 7;

– third, the Weibull scale factors determined using the CERAM computer code are smaller than those obtained using analytical equations. The discrepancy in scale factors reflects the accuracy of failure predictions.

	Method	Model	3-pt bending	4-pt bending	Biaxial flexure
Shape parameter m	Linear regression		23.4	23.8	22
Scale factor (MPa)	Linear regression	Weibull σ_o Batdorf		242 247	
	CERAM	Weibull σ_o MESM S_o	212.5 207	211 205	218 208

Table 9.3. *Statistical parameters for an alumina ceramic estimated from empirical fracture stress distributions*

Loading mode	σ_{ref} (MPa)	2 F_{max} (N)	P_{max} (MPa)
3-pt bending	400	260.41	
4-pt bending	350	280.45	
Biaxial flexure	375		13.307

Table 9.4. *Values of stresses used to correlate P_{CERAM} and P_{EXP} for the estimation of scale factors using equation [7.15]. $2F_{max}$ and P_{max} refer to the forces and the pressure used for the computation of these stresses and failure probabilities P_{CERAM}*

– *Failure predictions using the CERAM computer code*

Values of failure probability P_{CERAM} for bars and disks have been computed using the statistical parameters estimated from the results of 4-point bending tests (Table 9.3) for various values of force (3-point bending) and pressure (disks). Figure 9.4 shows the corresponding distributions of strengths, which fit quite nicely the empirical ones, which shows that the multiaxial elemental strength model leads to sound failure predictions. The trend in strength scaling agrees with the loading mode dependence of scale parameter values. Thus, the scale factor estimates for the multiaxial strength model showed quite no dependence on loading mode and failure strength scaling to uniaxial and biaxial stress states was satisfactory. The Weibull scale factors estimated using the CERAM computer code display discrepancy, as well as failure predictions shown in Figure 9.2.

9.2.2. *Unimodal population of internal defects*

Bending has been generated on samples of silicon carbide with a different form:

– on the one hand, bars subjected to 4- point bending;

– on the other hand, C-rings subjected to diametral compression (Figure 9.5).

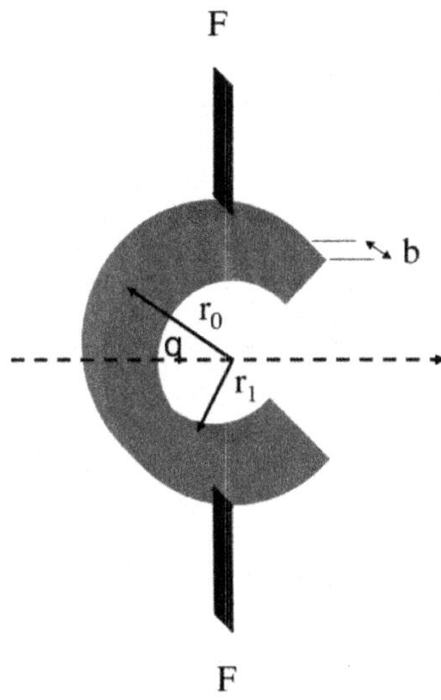

Figure 9.5. *Notations and symbols for the diametral compression of C ring*

This case study has been treated by Ferber *et al.* [FER 86]. They conducted tests and developed an analytical solution to calculate the C-ring strengths from 4-points bending data, using the uniaxial Weibull model. The test specimens were prepared out of tubes. Fracture was caused by internal defects, as shown by fractographic analysis of specimens. The experimental conditions are summarized in Table 9.5.

Loading mode	Specimens	Dimensions (mm)	Test conditions
4-pt bending	beams	2d = 3.1 b = 2.6 L =20.3	$2\ l_2 = 6.5$ mm $2\ l_1 = 19.05$ mm
Diametral compresion	C-rings	$r_o = 12.5$ $r_i = 9.3$ b = 5.2	

r_o = outer radius
r_i = inner radius

Table 9.5. *Conditions for the tests on silicon carbide specimens [FER 86]*

9.2.2.1. *Empirical distributions of failure stresses*

The empirical statistical distributions of rupture stresses were established by Ferber *et al.* [FER 86]. The failure stresses are derived from the values of force at specimen fracture using the theoretical expressions (equations [9.1] and [9.8]) of reference stresses (maximum stress in the specimen):

$$\sigma_{ref} = \frac{F_{max}}{A} \frac{R(r_o - r_a)}{r_o(R - r_a)} \text{ (diametral compression of C-Rings)} \quad [9.8]$$

with $r_a = \frac{1}{2}(r_o + r_i)$, $R = \frac{r_o - r_i}{Ln(r_o/r_i)}$, $A = b(r_0 - r_i)$.

The symbols are defined in Table 9.5 and Figure 9.5. Failure probabilities were calculated using the estimator $P_i = \frac{i - 0.5}{n}$. The empirical statistical distributions of failure stresses are shown in Figure 9.6, as Weibull plots. They fit satisfactorily single lines that are quite parallel, which suggests that a single population of defects was responsible for fracture.

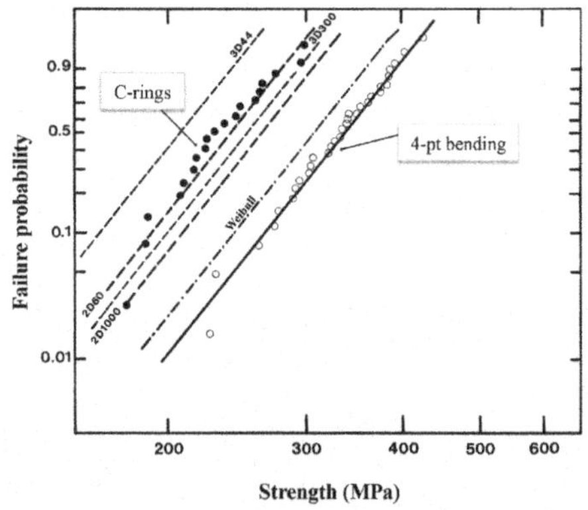

Figure 9.6. *Comparison of experimental C ring failure stress distribution established by Ferber et al. [FER 86] with predictions using analytical Weibull approach (by Ferber) and using the CERAM computer code (multiaxial elemental strength model, 2D and 3D analyses for 44, 60, 300 and 1,000 element meshes). The statistical parameters were estimated from the empirical distribution of 4-point bending failure stresses. The solid line represents experimental base line for predictions*

9.2.2.2. *Analytical calculation of failure probability [FER 86]*

The stress state in the C-rings is defined by the expressions for bending of curved beams [TIM 70]. It is a uniaxial flexural field including compressive components in the inner half, and tensile components in the outer half of C-ring. These two areas are separated by a neutral surface that is concentric of the internal and external faces. The expression of stresses is:

$$\sigma_\theta = \frac{F}{A}\frac{R(r-r_a)}{r(R-r_a)}\cos\theta \qquad [9.9]$$

where F is the compressive force applied to the ring, θ is the angle with the symmetry plane (Figure 9.5) and r is the distance from the center of the C-ring. The other symbols are defined in section 9.2.2.

The expression of failure probability is derived from the general Weibull equation [2.17] for a uniaxial stress state:

$$P = 1 - \exp\left[-\left(\frac{\sigma_{ref}}{\sigma_o}\right)^m b r_a^2 I_\theta I_r\right] \qquad [9.10]$$

where $I_\theta = 2\int_o^{\pi/2} \cos^m\theta d\theta$ and $I_r = \int_1^{ra/ro} \left(\frac{1-1/x}{1-r_a/r_o}\right)^m x dx$

The calculation of the probability of failure in 4-point bending is more trivial. Expressions are given in Chapter 5. The statistical parameters were estimated by linear regression analysis of a Weibull plot of failure stresses. They are given in Table 9.6. It may be noted that the values of Weibull scale factors (σ_o) are not invariant as would be expected in the presence of a single population of flaws, whereas the Weibull moduli are very close.

The failure probability of C-rings was calculated for the values of statistical parameters obtained from the 4 points bending results. Figure 9.6 shows that fracture was misestimated: the calculated strengths are significantly larger than the experimental values.

	Method	Model	4-pt bending	C-ring compression
Shape parameter m	Linear regression		7.9	8.0
Scale factor (MPa)	Linear regression	Weibull σ_o	122	101
	CERAM 2D (60 elements)	Weibull σ_o	29.7	29
		MESM S_o	25	25

MESM = Multiaxial elemental strength model

Table 9.6. *Statistical parameters for silicon carbide ceramic estimated from empirical fracture stress distributions*

9.2.2.3. *Numerical computation of failure probability*

– *Finite element analysis of stresses*

For symmetry reasons, a planar analysis in vertical midplanes of half specimens was performed. A mesh of 360 quadrangular elements was constructed for the 4-point bending conditions. Meshes of the C-ring comprised various numbers of two-dimensional (2D) or three-dimensional (3D) elements: 60 and 1,000 quadrangular elements, 44 and 300 3D elements. The following elastic properties of silicon carbide were introduced: Young's modulus E = 410 GPa, Poisson coefficient $\nu = 0.142$.

– *Estimation of statistical parameters*

The scale factor of each strength distribution has been estimated using the method discussed in Chapter 7, section 7.2.4. Probability of volume-originated fracture has been computed using the CERAM computer code for the shape parameters estimated by Ferber *et al.* (Table 9.6) for an arbitrary value of scale factor (taken to be the Weibull one estimated using analytical equations of failure probability, and given in Table 9.6), and for particular values of applied force. The corresponding values of σ_{ref} that have been computed using finite element method are given in Table 9.6. They are related to empirical values of failure probability P_{exp} through the empirical distribution of strengths. The values of scale factors have then been determined using equation [7.15].

It may be noted that estimates of scale factors σ_0 and S_0 are quite independent on loading mode. As previously, the scale factor of the multiaxial elemental strength model (S_0) is smaller than that of the Weibull model (σ_0). The σ_0 values are significantly smaller than the estimations by linear regression method obtained by Ferber *et al.* The invariance of scale factor determined using CERAM suggests that failure predictions will be satisfactory. Furthermore, it is consistent with the presence of a single population of fracture-inducing flaws.

– *Failure predictions using the CERAM computer code*

The failure probability of C-rings was computed for the statistical parameters obtained from 4-point bending results using CERAM (Table 9.6). Figure 9.7 shows the distribution of values of failure probabilities of mesh elements. Statistical distributions of C-ring failure stresses were established using values of maximum stress and corresponding specimen failure probabilities computed for various applied forces (Figure 9.6). They satisfactorily fit the empirical Weibull plot produced by Ferber. It may be noted that predictions depend slightly on mesh.

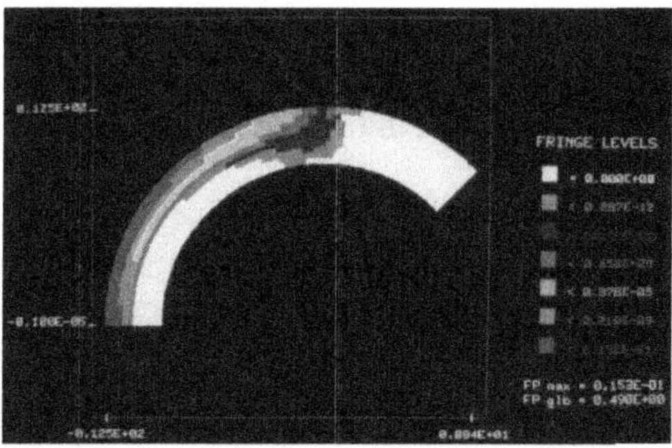

Figure 9.7. *Distribution of fracture probabilities in a silicon carbide C-ring under a diametral compression force of 250 N, computed using CERAM. For a color version of the figure, see www.iste.co.uk/lamon/brittle.zip*

The mesh firstly influences the computed stresses, which are underestimated when it is coarse, and the elements are too large with respect to stress gradients. Table 9.7 shows the effect of mesh on C-ring rupture data computed for the same value of applied force. The values of σ_{max} and failure probability increase with the number of elements.

	Numerical analysis				Theory	
	2D 60	2D 1000	3D 44	3D 300		
σ_{max} (MPa)	220	260	175	219	263	
Fracture	0.14	0.49	0.03	0.1		MESM
probability	0.16	0.54	0.09	0.12		Weibull

MESM = Multiaxial elemental strength model

Table 9.7. *Influence of the number of mesh elements on computed failure probability for a C-ring under diametral compression of 250 N*

However, when considering the statistical distribution of C-ring rupture stresses, the three finest meshes lead to close results, whereas the 3D 44 mesh appears too coarse as it leads to significant underestimation of failure. Table 9.7 shows that predictions using the Weibull model were close to results of the multiaxial elemental strength model, which was expected from the invariance of the scale factors obtained using the CERAM computer code, and which is consistent with the presence of uniaxial stress states.

Calculations using the CERAM package proved again more accurate than the analytical solution. Ferber's statistical parameters seem to be misestimated, which strongly influences the probability of failure.

9.2.3. *Bimodal population of internal and surface-located flaws*

Less common 3-point bending conditions have been applied to silicon nitride bars in order to generate a uniaxial or a multi-axial stress state. This was achieved with diverse span lengths, which were selected on the basis that at least one span be long enough to induce a predominantly tensile stress state, whereas the third span be sufficiently short that appreciable shear stresses exist within the test section. The experimental conditions are detailed in the literature [LAM 85b, LAM 90] and summarized in Table 9.8.

Loading mode	Specimens	Dimensions (mm)	Test conditions	τ_{max}/σ_{max}
3-pt bending	beams	2d = 2 b = 4 2 l = 21	Long support span	0.04
		2d = 4 b = 4 2 l = 9	Intermediate support span length	0.18
		2d = 4 b = 4 2 l = 6	Short support span	0.27

Table 9.8. *Conditions for the tests on silicon nitride specimens [LAM 85b]. Also given are the values of ratios of maximum shear stress to maximum tensile stress*

9.2.3.1. *Empirical statistical distributions of failure stresses*

The empirical distributions of rupture stresses were established using the theoretical expression of the maximum stress (equation [9.2]), and probability estimator $P_i = \frac{i-0.5}{n}$.

The flaws that induced brittle fracture of the specimens were identified using a scanning electron microscope:

– the failure origins of the long span specimens were predominantly located at the surface, whereas three failures at the low strength extreme seemed to originate from internal flaws;

– the failure of the short span specimens initiated predominantly from internal flaws, whereas three failures at the high strength extreme originated from surface-located flaws;

– fracture of the intermediate span specimens was caused by a bimodal flaw population including internal flaws at the low strength extreme and surface-located flaws at the high strength extreme.

The Weibull plots of empirical statistical distributions of failure stresses are shown in Figure 9.8. They fit single lines for the long and short span lengths, and two lines for the intermediate ones, which is consistent with the dominance of either a single or a bimodal flaw population.

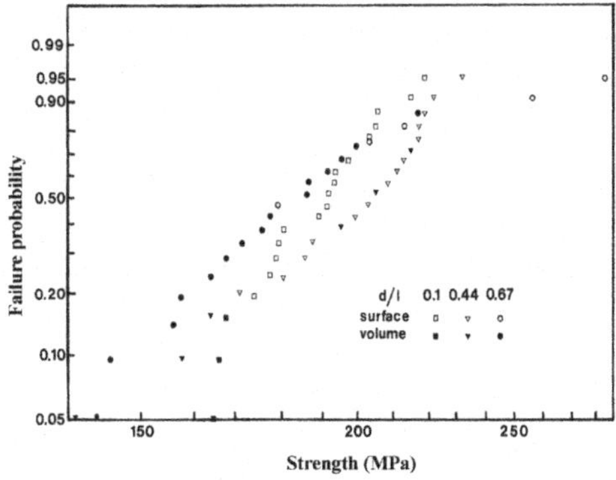

Figure 9.8. *Empirical distributions of failure stresses of silicon nitride bars established using 3-point bending tests with various support span lengths*

9.2.3.2. *Analytical calculation of failure probability*

The analytical solution is developed in the literature [LAM 85b]. The stress field consists of two shear components and a normal tensile component in the lower half of test specimens:

$$\sigma_z = \sigma_{max}\ (1-\frac{z}{l})(\frac{x}{d}) \qquad [9.11]$$

$$\tau_{xz} = \tau_{xz}^{max}\ (1-\frac{x^2}{d^2}) \qquad [9.12]$$

$$\tau_{yz} = \tau_{yz}^{max}\ \frac{x}{d}.\frac{y}{b} \qquad [9.13]$$

where $\tau_{xz}^{max} = \frac{1}{1+v}\frac{3F_{max}}{4bd}$, $\tau_{yz}^{max} = \frac{v}{1+v}\frac{3F_{max}}{2d^2}$

The expression of probability of failure from the internal flaws was derived from equation [4.54] of multiaxial elemental strength model. The Weibull equation [5.19] was applied to failures from the surface-located flaws in the long span specimens:

$$P_S = 1\text{-}\exp\left[-\left(\frac{\sigma}{\sigma_{os}}\right)^{m_s}\frac{2bl}{1+m_s}\right] \qquad [9.14]$$

$$P_V = 1 - \exp\left[-\left(\frac{\sigma}{\sigma_{ov}}\right)^{m_v} 2bdl \;\; I_V(m_v,0,0) \;\; H(m_v,\frac{d}{l})\right] \qquad [9.15]$$

where $H(m_v,\frac{d}{l}) = \int_V h^m dV$, with $h(\frac{x}{d},\frac{z}{l},\frac{d}{l}) = \frac{\sigma_1}{\sigma_{max}}$, σ_1 is a principal stress component.

The probability of failure from concurrent flaw populations in the intermediate span specimens was derived from equations [9.14] and [9.15] as:

$$P = 1 - (1 - P_S)(1 - P_V) \qquad [9.16]$$

The statistical parameters pertinent to internal and surface-located flaws were estimated by fitting the linearized forms of equations [9.14] and [9.15] to the appropriate Weibull plots. In the presence of single flaw populations, these are the Weibull plots of data obtained on short-span and long-span specimens. In the presence of concurrent flaw populations, the failure data were treated using the censored data method developed by Johnson and described in Chapter 7 (section 7.4.4). The data were reordered to conform with the two dominant populations. The statistical parameters pertinent to a population were estimated by fitting the linearized forms of equations [9.14] and [9.15] to the appropriate Weibull plots of the reordered data, omitting the censored data of the other population. An iterative least-squares fitting method was employed. The statistical parameters are summarized in Table 9.9. Note that the parameters of a population are quite invariant.

	Method	Model	Short support span	Intermediate support span	Long support span
Shape parameter m	Least squares		7.7	11.9 (surface) 7.6 (volume)	11.3
Scale factor (MPa)	Least squares CERAM	MESM S_o MESM S_o Weibull σ_o	11 11.3 12	72 (surface) 12 (volume)	69 59.5 63.5
			volume		surface

MESM = Multiaxial elemental strength model

Table 9.9. *Statistical parameters for silicon nitride ceramic estimated from empirical fracture stress distributions*

The calculation of failure probability for those test specimens with intermediate span was based on equation [9.16], for the statistical parameters of the two populations of defects estimated on lots of specimens with long and short spans. Figure 9.9 shows that rupture is underestimated: the calculated failure stresses are smaller than experimental ones.

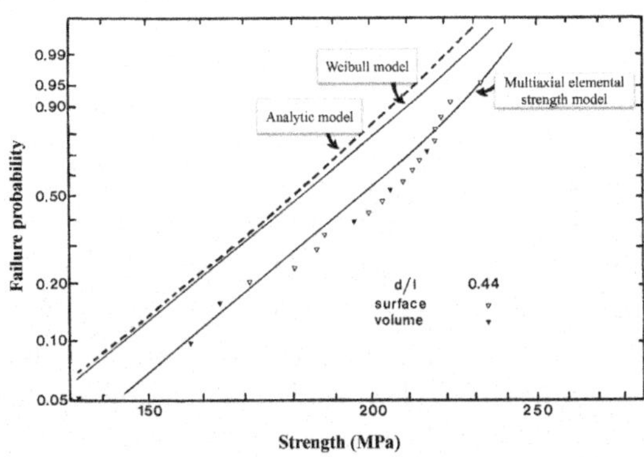

Figure 9.9. *Predictions of failure probabilities of beams during 3-point bending tests with intermediate support span length, using the Weibull analytical equation and the CERAM computer code (Weibull model and multiaxial elemental strength model)*

9.2.3.3. *Numerical computation of failure probability*

– *Finite element analysis of stresses*

For symmetry reasons, the analysis was performed in vertical longitudinal midplanes of half beams. The mesh was refined in the central area of test specimens, and it was coarser in the vicinity of supports to reduce the concentration of stresses. This stress concentration was particularly increased in the short span specimens. In this case, 60 elements are sufficient. The elastic properties of this silicon nitride ceramic are: E = 140 GPa, $\nu = 0.25$ [LAM 85b].

– *Estimation of statistical parameters*

The scale factors pertinent to surface-located and internal flaws were estimated using the CERAM computer code (as discussed in Chapter 7, section 7.1.4). Failure probabilities of long span and short span specimens were computed for arbitrary values of shape parameter (taken to be the values obtained using the analytical equations: Table 9.9), for the shape parameter values given in Table 9.9, and for particular values of applied force. The corresponding values of σ_{ref} that have been computed using finite element method are related to empirical values of failure probability P_{exp} through the empirical distribution of strengths. The values of scale factors have then been determined using equation [7.15] for P_{CERAM} and P_{exp} values related through σ_{ref}.

The estimates of scale factors are given in Table 9.9. For internal defects, they are smaller than those previously determined using analytical equations of failure probability. Furthermore, the general trend in scale factors is observed again: the scale factors of the multiaxial elemental strength model are smaller than the Weibull ones.

– *Failure predictions using the CERAM computer code*

The failure probability of intermediate span specimens was computed for the statistical parameters obtained from the failure data of long span and short span specimens using CERAM (Table 9.9). Statistical distributions of failure stresses were established using values of maximum stress and corresponding specimen failure probability computed for various applied forces. Figure 9.9 shows that the multiaxial elemental strength model predicts failures that are in excellent agreement with experimental results. By contrast, predictions based on the Weibull model underestimate failure strengths. This trend may appear as inconsistent with expectation. Since shear stresses are accounted for in the multiaxial elemental strength model, we could expect larger strength predictions by the Weibull model. This trend is related to large Weibull scale factor values [LAM 90].

It is also interesting to note that the computerized version of the multiaxial elemental strength model gives more precise results than the analytical approach. This discrepancy can be attributed to the following:

– the analytical approach is based on a stress state definition that is less precise than the results of finite element analysis, which accounted for all the stress components, particularly in the surface of beams;

– failure from the surface-located flaws was treated using the uniaxial Weibull model.

Numerical computations prove again to be more accurate than the analytical solutions. The multiaxial elemental stress model provides satisfactory predictions. Furthermore, the numerical approach is homogeneous, as it is applied to both the estimation of statistical parameters and predictions of failure.

9.3. Prediction of thermal shock failure

Thermal shock testing is an alternate deformation-controlled loading mode for which deformations are generated by transient temperature gradients. When a solid is placed in a medium at different temperature, heat exchanges occur that generate temperature gradients, and therefore a non-uniform deformation state. The resulting stresses evolve over time, reach a maximum locally as the temperature front progresses toward the interior of the solid. Failure probability depends on time. Equation [7.1] is expressed as:

$$P(t) = 1 - \exp\left[-\frac{V}{V_o} K(t) \left(\frac{\sigma_{ref}(t)}{\lambda_o} \right)^m \right] \qquad [9.17]$$

where $\sigma_{ref}(t)$ is a reference stress that is defined as the peak stress in specimen at time t, $\sigma_{ref}(t)$ is time dependent, as well as K(t) that depends on stress state.

The main difficulty in the calculation of failure probability in the presence of a transient stress field is to establish the expressions of the stress state necessary for the calculation of the K (t) function, which may partly explain the small number of works devoted to the application of probabilistic approaches to failure by thermal shock.

9.3.1. *Quenching of alumina disks [LAM 85, LAM 93]*

In [LAM 85a, LAM 93], thermal shocks were carried out on alumina disks (5 cm in diameter and 0.25 cm thick) using the device shown in Figure 9.10. The disk specimen is supported along a concentric line support near its periphery. It is heated up to a selected temperature and, then, cooled down abruptly within the furnace using high-velocity helium channeled onto the disk center. Depending on initial temperature, and also on disk tested, it may break or not. The initial temperature of unbroken or uncracked disks is increased by 10°C until fracture occurs during thermal shock. Thus, the critical applied temperature (T_c) at which disk fracture is observed can be determined to within 9°C.

9.3.1.1. *Empirical distribution of thermal failure data*

The resistance to thermal shock is measured by the differential between the initial disk temperature and the temperature of helium jet: at disk fracture $\Delta T_c = T_c - T_e$, where T_e is the temperature of helium jet (T_e = 10°C). Twenty-eight disks were tested. This procedure provides a set of thermal shock fracture data. The values of thermal resistance ΔT_c were handled as statistical data. Empirical statistical distribution was established according to the method described in Chapter 7. The Weibull plot of thermal strengths ΔT_c is shown in Figure 9.11. It is interesting to note that the data points are distributed on a single line, in accordance with the results of fractographic analysis of specimens after thermal shocks which indicated that the fracture origins belong to a population of surface-located defects.

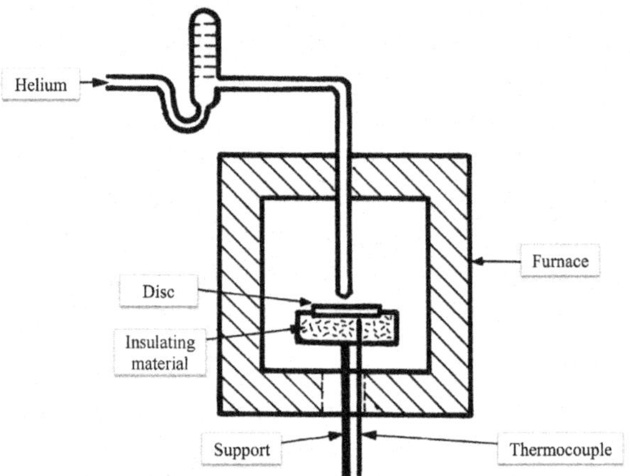

Figure 9.10. *Schematic of set up used for thermal shock quench tests on the alumina disks*

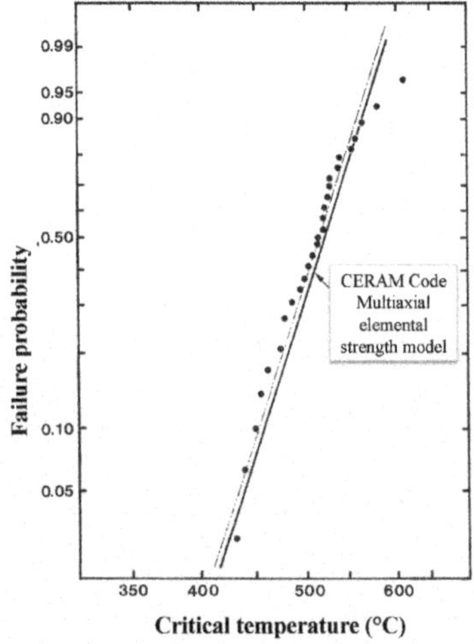

Figure 9.11. *Weibull plot of thermal strengths ΔT_C for alumina disks. Solid line represents CERAM prediction (multiaxial elemental strength model)*

9.3.1.2. *Estimation of flaw strength parameters*

Flaw strength parameters pertinent to the population of surface defects were estimated from strengths measured by biaxial disk flexure. The disk is supported by a concentric ring. The effort is applied at a constant rate (2 MPa sec-1) at the center through a flat end piston (Table 9.10). The stress field is axisymmetric. Within the area located below the piston, the stress state is uniform and equibiaxial on every horizontal plane. The peak stress is reached at the center of the outer surface of disk. The biaxial flexural strength can be estimated from the maximum force as follows [KIR 67]:

$$\sigma_{max} = \frac{3F_{max}}{4\pi e^2}\left[2(1+\nu)\mathrm{Ln}\frac{r_1}{r_3} + \frac{(1+\nu)(r_1^2-r_3^2)}{r_1^2}\frac{r_1^2}{r_2^2}\right] \qquad [9.18]$$

Loading mode	Specimens	Dimensions	Test conditions
Biaxial flexure	discs	$2\ r_2 = 50$ mm e = 2.5 mm	$2\ r_1 = 32.90$ mm support diameter $r_3 = 1.63$ mm loading piston radius
Thermal shocks	discs	$2\ r_2 = 50$ mm e = 2.5 mm	Quenching by helium jet (jet temperature = 10°C) Jet diameter 1.42 mm Flow rate 14 l/min

Table 9.10. *Conditions for the thermal shocks and the biaxial flexure tests on alumina disks [LAM 85a]*

Symbols are defined in Table 9.10. The empirical distribution of flexural strengths was established using the estimator $P_i = \frac{i-0,5}{n}$. The Weibull plot is shown in Figure 9.12. Most failures resulted from defects located in the outer surface of disks. The fractographic analysis of specimens has also identified a few fracture-inducing defects located below the surface.

The statistical parameters pertinent to surface-located flaws were estimated from the Weibull plot of the reordered data, omitting the censored data of the population of subsurface flaws according to the censored data method discussed in Chapter 7. The shape parameter was derived from the slope of the regression line. The scale factor was estimated using the multiaxial elemental strength model and CERAM. An axisymmetric 2D finite element analysis of stresses was carried out for a mesh of 313 quadrangular elements. The values of scale factors have then been determined using equation [7.15] for P_{CERAM} and P_{exp} values related through σ_{ref}, as in the previous case studies. The estimates of statistical parameters are summarized in Table 9.11. It is worth pointing out that the shape parameter is close to that estimated from the slope of the Weibull plot of thermal strengths ΔT_c.

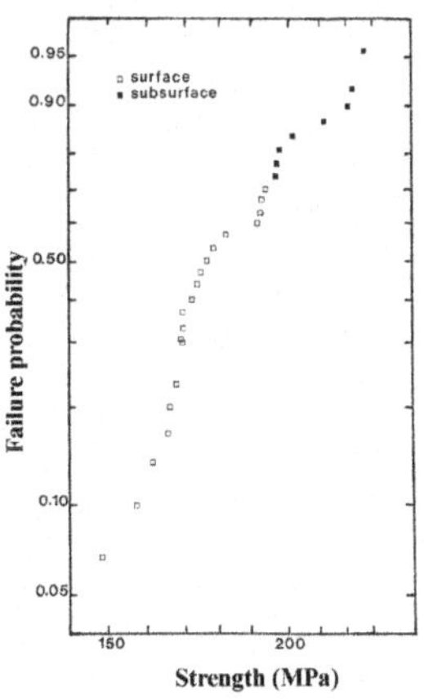

Figure 9.12. *Empirical distribution of alumina disk failure stresses determined using biaxial flexure at room temperature*

	Method	Model	Biaxial flexure	Thermal shocks
Shape parameter m	Linear regression		14	13.2
Shape parameter (MPa)	CERAM	MESM S_0	73	

MESM = Multiaxial elemental strength model

Table 9.11. *Statistical parameters for an alumina ceramic estimated from empirical distributions of fracture stresses measured in biaxial flexure at room temperature*

9.3.1.3. *Predictions of thermal shock failure*

– *Finite element analysis of thermal stresses*

The thermal stresses generated during thermal shocks were computed using the finite element method in an axisymmetric cell corresponding to half disk (Figure 9.13). The mesh consisting of 313 elements is shown in Figure 9.13.

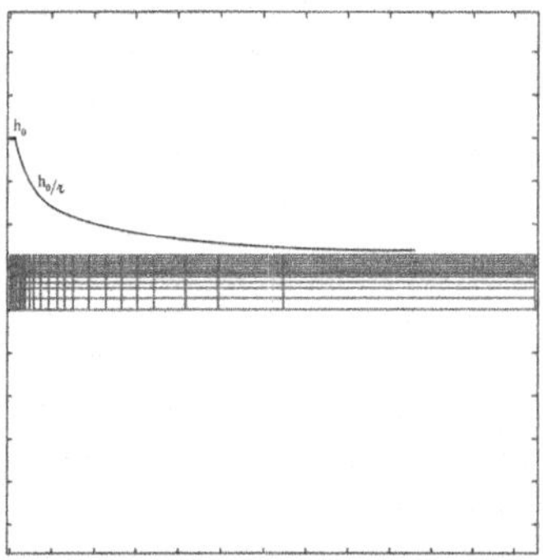

Figure 9.13. *Finite element mesh for the analysis of temperature and stress fields operating on a disk during a quench test. The thermal boundary conditions are specified through heat transfer profile*

Distributions of temperatures within disks were calculated at various time steps (2 s), using the TAURUS computer program. The boundary conditions replicate the heat exchanges between the disk and the environment during cooling. Convective heat transfer was considered through the upper surface. The heat transfer factor coefficient (h) through the surface was assumed to be uniform over the diameter of helium jet (1.52 mm; Table 9.10) and to decrease inversely with distance from the jet circumference to account for flow attenuation:

$$h = 0.256 \text{ (W/cm}^2\text{ sec) for } r \leq 0.76 \text{ mm} \qquad [9.19]$$

$$h = \frac{0.256}{r} \text{ (W/cm}^2\text{ sec) for } r > 0.76 \text{ mm}$$

where r is the distance from the center of disk.

The thermoelastic properties of alumina ceramic are given in Table 9.12.

Properties	Al_2O_3
Density (g/cm^3)	3.9
Young's modulus (GPa)	390
Thermal expansion coefficient (°C^{-1})	8.3 10^{-6}
Heat capacity (J/kg.K)	1200
Thermal conductivity (W/m.K)	11.5

Table 9.12. *Thermoelastic properties of the alumina ceramic subjected to thermal shocks*

Figure 9.14 shows the thermal stress profiles in the outer surface of disk during cooling. Like in biaxial flexure, the stress is equibiaxial and it is maximum at the disk center. The stresses on the surface increase during cooling. Figure 9.15 illustrates the evolution of the maximum stresses at the center of disk. They reach a maximum after 12 s, and then decrease.

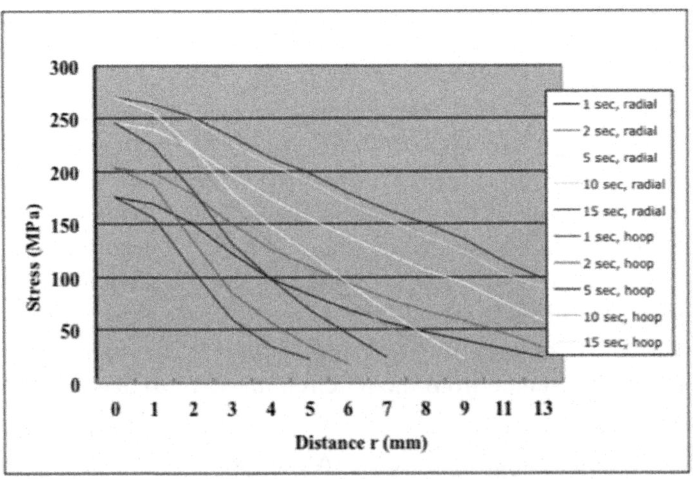

Figure 9.14. *Evolution of profiles of radial and hoop stress components in the quenched surface of disk during cooling: initial temperature of disk = 500°C. For a color version of the figure, see www.iste.co.uk/lamon/brittle.zip*

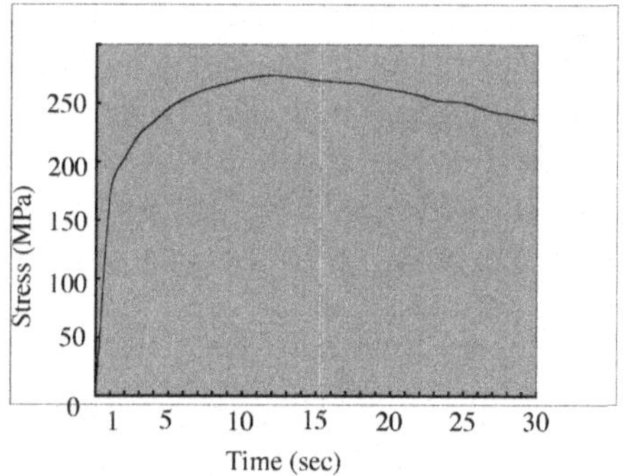

Figure 9.15. *Evolution of maximum stress at the center of disk surface exposed to the helium jet (r = 0, $\sigma_{rr} = \sigma_{\theta\theta}$) during a quench test: initial temperature of disk = 500°C*

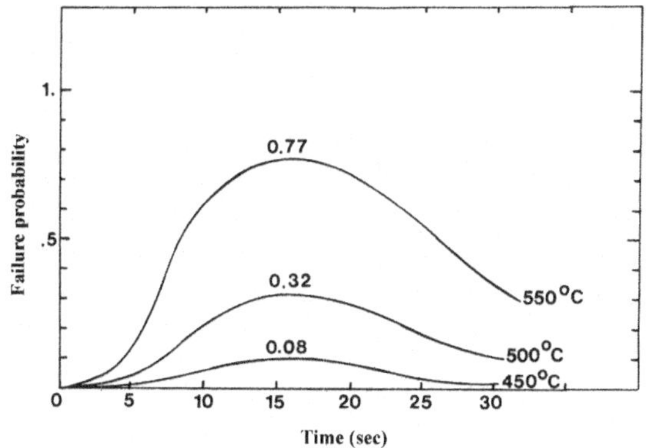

Figure 9.16. *Evolution of failure probability during quench tests on alumina disks having various initial temperatures (computed using CERAM and the multiaxial elemental strength model)*

– *Computation of failure probability*

The probability of disk failure during thermal shocks was computed using CERAM for various applied temperature differentials and for the flaw strength statistical parameters given in Table 9.11. Figure 9.16 shows that the probability of failure during a thermal shock increases, reaches a maximum after 15 s and then decreases. This result reflects observations during thermal shock tests since cracking was evidenced by a typical audible burst that was heard a few seconds after the beginning of the quenching. This latent period has been reported by Manson and Smith who applied Weibull's statistical theory to the thermal shock problem. Their computations showed that the time of maximum risk of rupture does not coincide with initiation of the quenching.

The rupture occurs at the moment when one of the flaws becomes critical. In the presence of a transient stress state, this requirement is met when a sufficient amount of material is operated on by stresses sufficiently high with respect to the severity of flaw population. The maximum of failure probability indicates the fraction of disk broken at each applied temperature, that is the value of probability that can be associated with each critical temperature difference to establish the statistical distribution of thermal shock resistance ΔTc. Figure 9.11 shows that the thermal shock failure probability was satisfactorily predicted. This example demonstrates that the thermal shock fracture can be predicted by calculations of probability of failure, which is in line with the notion of parity between thermal shock failure and mechanical failure in the absence of thermally induced effects. It is necessary to perform appropriate tests to estimate the flaw strength parameters. In this case of disk quenching, disk flexure testing at room temperature turned out to be a sound approach, as it allowed activation of fracture-inducing flaws under biaxial stress state. However, thermal shock tests could have been used too. To this end, fracture characteristics (such as critical temperature) that can be measured must be defined and related to the stress state so that calculations of the probability of failure can be made and therefore the flaw strength parameters derived from thermal shock tests. Furthermore, computation of the stress field by the finite element method requires that the boundary conditions reproduce as faithfully as possible the heat transfer conditions.

9.3.2. *Thermal fatigue [LAM 91]*

Repeated thermal shock heat ups were produced using a plasma jet [LAM 91]. This testing technique makes it possible to generate combined cyclic and mixed-mode loading. Bend bars of silicon nitride ceramics were mounted on a roundabout holder rotating at constant speed (Figure 9.17). At each turn, the front face of a sample was impinged by the jet for about 2 s, and temperature rose to about 1,200°C (Figure 9.18). The impinged area was about 20 mm long. Then, the sample was cooling down out of the jet for approximately 4 s in ambient air, that is the time needed to complete the turn of roundabout (Table 9.13). Batches of 20 specimens were subjected to various numbers of cycles.

Loading mode	Specimens	Dimensions	Test conditions
4-pt bending	beams	2d = 3 mm b = 5 mm L =50 mm	$2\,l_2$ = 20 mm $2\,l_1$ = 40 mm
Repeated heat-ups	beams	2d = 3 mm b = 5 mm L =50 mm	Heat up by plasma jet (2 seconds) Cooling down in air (4 seconds) 10 laps / minute

Table 9.13. *Conditions for the thermal shock heat ups and the 4-point bending tests on silicon nitride beams [LAM 91]*

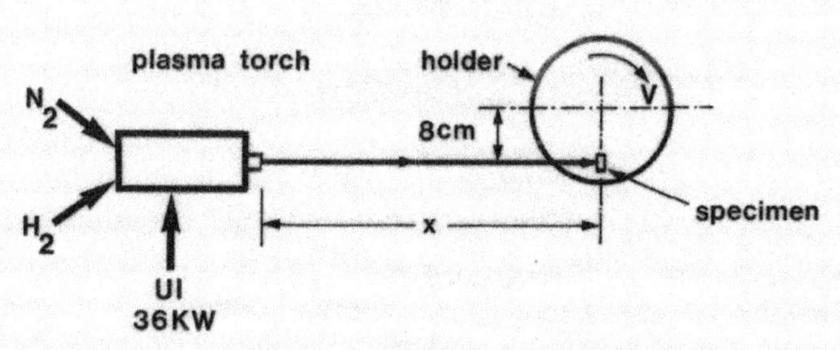

Figure 9.17. *Sketch showing a top view of the plasma jet facility*

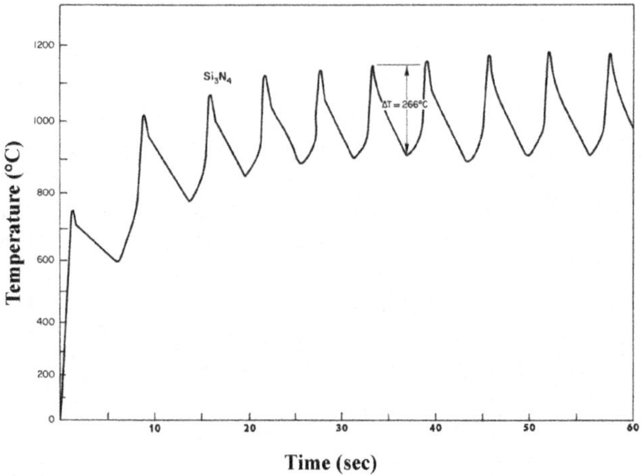

Figure 9.18. *Temperature cycles measured at the surface of a test specimen during repeated thermal shock heat ups*

Properties	Si_3N_4	Si_3N_4-ZT
Density	3.14	2.45
Young's modulus (GPa)	280	165
Strength (MPa)	500	200
Thermal expansion coefficient (°C^{-1})	3 10^{-6}	3 10^{-6}
Heat capacity (J/kg °K)	1000	1000
Thermal conductivity (W/m°K)	15	1.5

Table 9.14. *Properties of the silicon nitride ceramics subjected to repeated thermal heat ups*

The characteristics of the two silicon nitride ceramics that were tested are given in Table 9.14. The so-called Si_3N_4-ZT displayed low properties: low strength, density and thermal conductivity. The Si_3N_4-ZT samples were broken after a few cycles. The S_3N_4 samples survived more than thousand cycles, however, they suffered some degradation that was characterized by residual strengths. Initial and residual strengths were measured in 4-point bending at room temperature. Figure 9.19 shows the degradation of empirical strength distribution after large numbers of repeated thermal shock heat-ups. SEM fractography indicated that flexural fracture of samples having experienced no or less than 10 cycles was caused by populations of internal and surface-located voids. Microcracked regions were also identified after great numbers of cycles. Failure probabilities during the first three cycles were calculated using the post-processor CERAM for both Si_3N_4 ceramics.

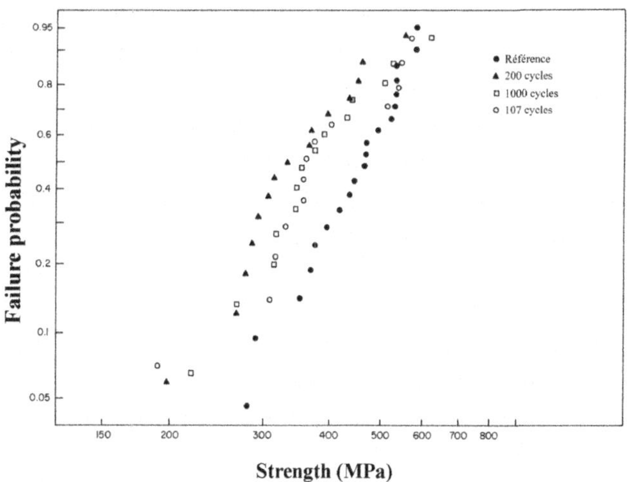

Figure 9.19. *Empirical statistical distributions of initial and residual strengths after large numbers of repeated heat ups, measured using 4-point bending tests at room temperature on Si_3N_4 samples*

9.3.2.1. *Estimation of statistical parameters*

The empirical distributions of rupture stresses were established from the failure data measured using 4-point bending tests at room temperature. The method employed is detailed in previous sections as well as the theoretical expression of failure stress (equation [9.2]). The data were reordered to conform with the presence of internal and surface-located flaw populations using the censored data method developed by Johnson and described in Chapter 7 (section 7.5.4). The statistical parameters pertinent to a population were estimated from the appropriate reordered data, omitting the censored data of the other population. The shape parameter m was estimated using linear regression analysis of Weibull plots constructed using probability estimator $P_i = \frac{i-0.5}{n}$. The scale factors were estimated using CERAM. An axisymmetric. 2D finite element analysis was carried out for a mesh of 125 quadrangular elements similar to that used for the thermal shock failure predictions. The values of scale factors have then been determined using equation [7.15] for P_{CERAM} and P_{exp} values related through σ_{ref}, as in the previous case studies. The estimates of statistical parameters obtained on as-received specimens are summarized in Table 9.15. As previously mentioned, the scale factors of the multiaxial elemental strength model are smaller than the Weibull ones.

9.3.2.2. *Computation of thermal stresses*

A two-dimensional analysis was carried out on half specimen for symmetry reasons. A regular mesh of 125 quadrangular elements was constructed. The specified boundary conditions reproduce the applied thermal loading described by the temperature cycles measured at the surface of bars as shown in Figure 9.18. A uniform temperature was specified over the impinged area. Thermal and mechanical properties assumed for the analysis are given in Table 9.14. Examples of temperature distributions in specimens computed at different time steps are shown in Figure 9.20. They show the extension of the hot zone within the bar.

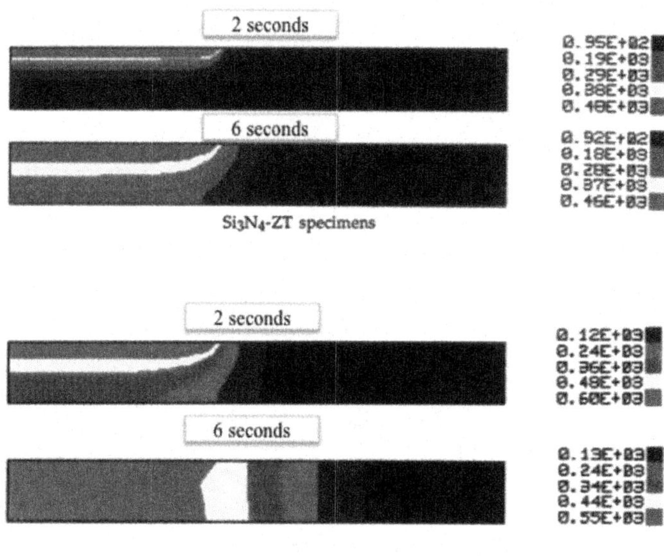

Figure 9.20. *Computed distributions of temperatures in the Si_3N_4 and Si_3N_4-ZT specimens after the first heating (at time 2 s) and the subsequent cooling (at time 6 s). Half the specimen was analyzed for symmetry reasons. Heat transfer was specified through the upper face. For a color version of the figure, see www.iste.co.uk/lamon/brittle.zip*

Stress analysis shows that the thermally induced stress field includes tensile and shear components. The peak tensile and shear stresses have comparable magnitudes. The higher stresses are observed on the low conductivity Si_3N_4-ZT specimens. At each cycle, the impinged and the rear faces are subjected alternately to tensile or compressive stresses combined with shear components. Tensile stresses arise in the impinged region on cooling, and then in the rear region on heating, except at the first heat up when they operate in the midplane.

9.3.2.3. *Prediction of failure*

Failure probabilities during the first three thermal cycles were then computed using the CERAM computer code, the multiaxial elemental strength model and the pertinent flaw strength parameters given in Table 9.15.

	Method	Model	4-pt bending Si_3N_4	4-pt bending Si_3N_4-ZT
Shape parameter m	Linear regression		5.2	5.2
Scale factor (MPa)	CERAM 2D	Weibull σ_o	Surface: 71.4	32.2
			Volume: 17	7.7
		MESM S	Surface: 57.6	26
			Volume: 13.7	6.2

MESM = Multiaxial elemental strength model

Table 9.15. *Statistical parameters for silicon nitride ceramics estimated from empirical distributions of 4-point bending fracture stresses*

Distributions of failure probabilities in the low-conductivity Si_3N_4-ZT samples display a region of high failure probability, located in the midplane at the first heat up, then alternately in the impinged (on cooling) or in the rear (on heating) parts at subsequent cycles (Figure 9.21). In the high-conductivity Si_3N_4, the highest failure probabilities are observed in the midplane region during heat up. These distributions of failure probabilities are in agreement with the experimental behavior of specimens and also with the stress distributions. The high failure probability regions (in the midplane and in the cross-section) correspond to the location of failure sites in the Si_3N_4-ZT specimens.

Values obtained for specimen failure probabilities are given in Table 9.16. Failure probability of Si_3N_4-ZT specimens reaches 1 at the third heat up, which indicates that fracture should occur. It is 0.35 at the first cycle, which indicates that 35% of specimens should fail at the first cycle.

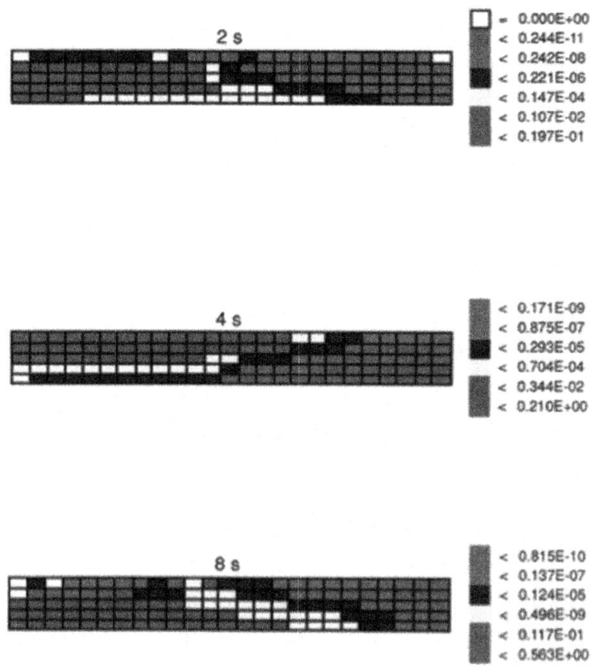

Figure 9.21. *Distributions of failure probabilities in the Si_3N_4-ZT specimens during plasma jet thermal fatigue: during the first and second heat ups (at times 2 and 8 s) and the first cooling (at time 4 s). Half the specimen was analyzed for symmetry reasons. Heat transfer was specified through the upper face. For a color version of the figure, see www.iste.co.uk/lamon/brittle.zip*

Cycle	Time (seconds)	Computed failure probability Si_3N_4-ZT	Si_3N_4
1	2	0.35	$0.16\ 10^{-4}$
	4	0.75	$0.6\ 10^{-6}$
	6	$0.4\ 10^{-3}$	$0.83\ 10^{-8}$
2	8	0.998	$0.10\ 10^{-5}$
	10	1	$0.79\ 10^{-7}$
	12	0.07	$0.69\ 10^{-8}$
3	14	1	$0.23 10^{-6}$

Table 9.16. *Probabilities of thermal shock fracture computed using the CERAM computer code (multiaxial elemental strength model)*

By contrast, the failure probability of Si_3N_4 specimens is very low ($< 10^{-4}$). It is unlikely during the first cycles. Furthermore, the highest failure probabilities in specimens do not reach high values. Failure probabilities at larger numbers of cycles can be computed for appropriate values of flaw strength parameters that decrease as indicated by the residual strength distributions (Figure 9.19). Results would show whether failure probability increases to significant values that would give amounts of broken specimens.

Predictions of high failure probability for the Si_3N_4-ZT specimens can be interpreted on the basis of poor properties of Si_3N_4-ZT. Si_3N_4-ZT is subject to high thermally induced stresses as a result of low conductivity and it possesses small tensile strength.

Thus, the trends in thermal shock behavior that can be qualitatively anticipated from stress distributions and material respective properties are quantitatively predicted by the failure probability computations. Due to the complexity and polyaxiality of stress state, numerical computations using the multiaxial elemental strength model are required.

9.4. Conclusion

Computations of failure probability provide a powerful tool for failure predictions. They can be employed in numerous loading cases, provided that the boundary conditions can be replicated precisely, and the flaw strength parameters can be estimated for all the active flaw populations under stresses. Therefore, various test conditions may be required for the establishment of empirical failure strength distributions. Thus, flexural loading modes offer a wide applicability.

However, when the stress state is defined approximately by analytical expressions, numerical analysis provides more reliable failure predictions. For this purpose, it is required that we determine the statistical parameters also using numerical computations of stress state and failure probability. When the evolution of flaws due to environmental effect is accounted for by time-dependent flaw strength parameters, time-dependent failure probability indicates a lifetime when it reaches maximum value of 1. Delayed failure does not result only in slow growth of a single crack. It may be caused by the evolution of populations of fracture-inducing flaws. It may also result from the action of transient stress state, like those generated by thermal shocks.

Analytical analyses of failure probability often produce erroneous results. Many examples can be found in the literature. The multiaxial elemental strength model provides satisfactory failure predictions. However, the finite element mesh must optimized. The flaw strength parameters of the multiaxial elemental strength model are smaller than the Weibull statistical parameters.

10

Application of Statistical-Probabilistic Approaches to Damage and Fracture of Composite Materials and Structures

10.1. Introduction

The previous chapters addressed brittle fracture, which displays the following features: (1) the initiation of a crack from a defect causes the instantaneous breakdown of a solid, (2) crack initiation is the preponderant step as it coincides with the complete separation of the solid, (3) the crack propagates instantaneously through the solid and (4) the material does not generate crack arrest mechanism. As a result, the most severe flaw is responsible for fracture. The flaw density function is defined by a power law which approximates the low extreme of the flaw strength density function.

Those materials that can be assimilated to structures including fiber reinforced composites or multilayered materials are designed to develop mechanisms of crack arrest. In these structures made of an assembly of brittle elements like fibers, a matrix or multilayers, cracks initiate also from flaws, propagate through the element and are arrested at the closest interface. From this point of view, it may be considered that cracking of elements of the structure does not violate the weakest link concept and that it may be approached using extreme value theory-based models.

When the load is increased, the above mechanism of cracking repeats. Additional cracks can form from other flaws, which are less severe than the previous ones, but these flaws are the most severe in the population of remaining flaws. It is clear that models based on the extreme value theory do not apply when we consider the initial population of flaws and the whole structure. The fundamental approach based on the probability density function of flaw strengths (g(S)) can be extended by introducing the changes in the density function resulting from the elimination of flaws during the formation of previous cracks. This model requires an appropriate expression of g(S) with respect to the whole population of flaws that can be activated by stresses.

Models based on the extreme value theory can be devised when a network of cracks appears which delineate fragments of material. Cracking is thus regarded as the brittle fracture of a fragment from the most severe flaw present in the fragment. It is generally assumed that the expression of flaw strength density function is independent of volume, so that the size effects induced by the formation of successive can be described.

In a parallel approach, cracking is assimilated to the emergence of critical defects according to a Poisson process. Defects are considered to be punctual and their resistance is measured by a stress. This approach has been developed for the fragmentation of fibers in organic or metallic matrix composites.

Damage by multiple cracking and fragmentation is observed in a variety of materials made of at least one brittle constituent: fragmentation of the ceramic matrix and multiple fracture of tows in continuous fiber reinforced composites, fragmentation of fibers in polymer or metallic matrix composites reinforced by continuous inorganic fibers, fragmentation of layers in multilayered materials.

The chapter discusses the extension of statistical probabilistic approaches to brittle fracture to multiple cracking and fragmentation in composites. It proposes a way to treat problems posed by the behavior of modern materials made up of assembled constituents. It lays the foundations of a general theory of fracture that takes into account the causes of fracture or damage, as opposed to the classical macroscopic approaches that focus on symptoms.

10.2. Damage mode by successive cracking in continuous fiber reinforced composites

Examples of damage by successive microcracks observed in continuous fiber reinforced composites are shown in Figure 10.1. The microcracks form in one of the constituents having contrasting mechanical properties: in the fibers of unidirectional polymer matrix composites, in the ceramic matrix of unidirectional composites, in the transverse plies of $(0°/90°)_n$ laminates or in the transverse tows of woven composites. The microcracks are arrested at an interface between either the fiber and the matrix in a unidirectionally reinforced composite, or between crossing plies or tows in a two-dimensional (2D) composite. This damage process requires that the load is applied increasingly and that the cracked constituent is loaded. It stops when further loading of a fragment is no longer possible although the applied load is increased (for instance because the interface is unable to transfer load), or when the strength of fragments becomes larger than the operating stress as a result of size effect. The ultimate failure of composite material occurs when the population of fibers is no longer able to carry the applied load. The population of fibers in plies or bundles suffers damage prior to ultimate failure. As the fibers fail, either the load is transferred by the polymer matrix to neighboring fibers (local load sharing), or it is carried out by the surviving fibers only in ceramic matrix composites after saturation of matrix cracking (global load sharing).

Matrix cracking and fiber failure in composites can be modeled as a series system (matrix or fiber fragmentation) or a parallel system (damage of fiber bundles or matrix fragmentation). The damage in series system displays the following characteristics:

– the total volume under stress is constant during successive cracking;

– the fragments (Figure 10.1) are subject to stresses;

– the whole population of flaws present in the total volume of material is able to induce fracture;

– flaws are eliminated when new cracks form. Several flaws may be eliminated simultaneously when the stress state within fragments is affected by fiber matrix debonding.

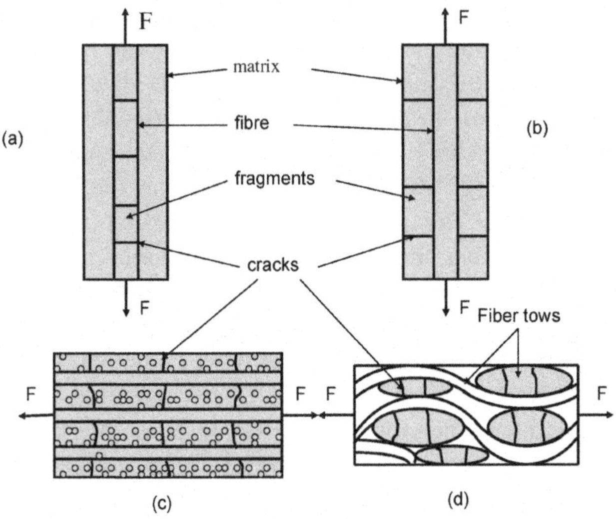

Figure 10.1. *Diagrams of damage modes by successive failures (series systems): a) fragmentation of the fiber in an organic matrix composite, b) fragmentation of the matrix in a single fiber reinforced ceramic matrix composite, c) multiple cracking of transverse plies in a laminated composite and (d) multiple cracking in a composite with woven plies*

By contrast, damage in the parallel system of independent elements (Figure 10.2) shows the following characteristics:

– the total volume under stress decreases as elements break;

– a broken element is unloaded after it fractured;

– the fracture of an element is caused by the most severe flaw present in that element. The population of fracture-inducing flaws is restricted to the low strength extreme of the whole population of flaws;

– the fracture of an element obeys the weakest link concept, and the statistical-probabilistic models discussed in previous chapters.

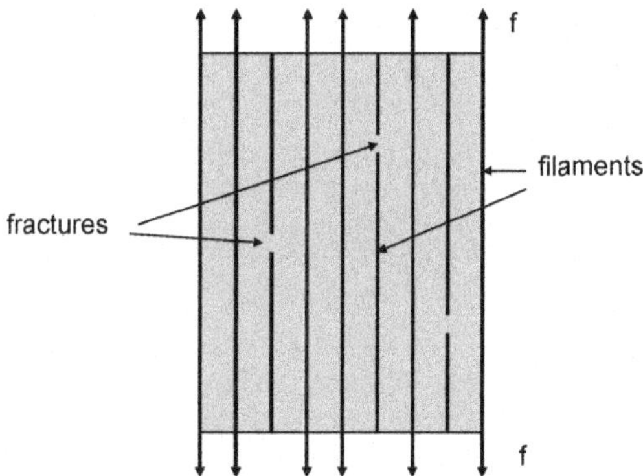

Figure 10.2. *Damage by filament fractures to a dry tow of parallel and independent filaments (system of parallel elements) under uniaxial tensile forces (f) on filaments*

10.3. Flaw populations involved in damage and pertinent flaw strength density functions

During damage by successive cracking, a significant fraction of the whole population of flaws is critical, as opposed to brittle fracture that involves the weakest flaw. As a result, the range of corresponding elemental strengths is broader. However, the larger elemental strength S_{max} is smaller than the theoretical strength (also termed ideal cohesive strength) that is required to break the interatomic links in the fracture plane (equation [1.1] Chapter 1). In this presentation, it is assumed that the flaw density function g(S) can be described by a Weibull function on the total range of elemental strengths [S_{min}, S_{max}] (Figure 10.3):

$$g(S) = m\frac{S^{m-1}}{S_o^m}\exp\left[-(\frac{S}{S^0})m\right] \qquad [10.1]$$

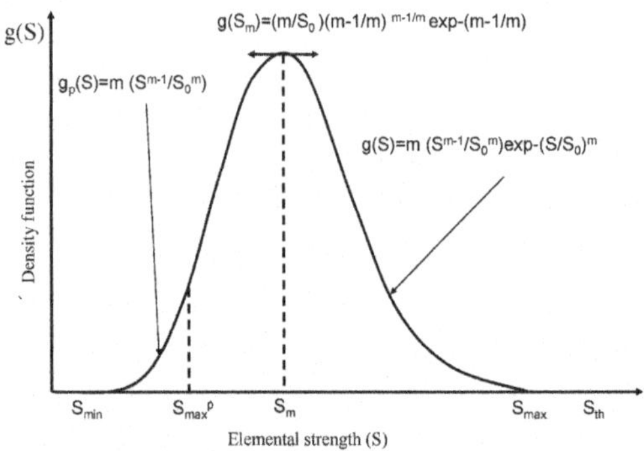

Figure 10.3. *Flaw strength density function g(S) defined by the Weibull expression*

It should be recalled first, that m and S_o refer to the flaw strengths that is to elemental strengths of volume elements of identical size containing each a single flaw. The values of m and S_o depend on the range of elemental strengths, as shown by the expressions used in the maximum likelihood technique (Chapter 7). It is anticipated that S_o increases and m decreases with increasing S_{max} (S_{min} being constant). Furthermore, these values of m and S_o are different from those estimated from the brittle failure of test specimens that would correspond to a smaller range of strength data at the low strength extreme.

For small values of S at the low strength extreme, the expression [10.1] becomes:

$$g(S) = m\frac{S^{m-1}}{S_o^m} \qquad [10.2]$$

The difference between values of flaw density functions given by equations [10.1] and [10.2] is given by equation [10.3]. It increases with S:

$$\Delta g(S) = m\frac{S^{m-1}}{S_o^m}\left[1 - \exp\left[-(\frac{S}{S_o})m\right]\right] \qquad [10.3]$$

The error when using equation [10.2] can be determined using equation [10.3], provided that the values of appropriate flaw strength parameters are available.

Expression [10.2] is generally introduced in statistical-probabilistic models of brittle failure, as discussed in previous chapters. In the following, the subscript p is used to denote brittle failure:

$$g_p(S) = m_p \frac{S^{mp-1}}{S_o^{mP}} \qquad [10.4]$$

Expression [10.4] corresponds to flaw strength range $[S_{min}, S^P_{max}]$ with S_{min} identical to above and $S^P_{max} < S_{max}$ (Figure 10.3). m_p and S_{op} designate the corresponding values of statistical parameters. The difference between $g_P(S)$ and $g(S)$ also increases with S, as indicated by equation [10.5]:

$$g_p(S) - g(S) = \frac{m_p S^{m_p - 1}}{S_{op}^{mp}} - \frac{m S^{m-1}}{S_o^{m-1}} \exp\left[-(\frac{S}{S_o})^m \right] \qquad [10.5]$$

The error when using $g_p(S)$ can be estimated from equation [10.5]. It is probably larger than Δg (s) since it is anticipated that $m_p > m$ and $S_{op} < S_o$.

The majority of works on the fragmentation of fibers are based on a power law similar to expression [10.2] with the statistical parameters estimated from results of brittle failure tests.

In summary:

– for the brittle fracture of samples tested separately (parallel system), the elemental strengths of critical defects cover an extreme of the density function g(S) of the entire flaw population. By contrast, a greater proportion of the whole of the flaw population causes successive failures in the damage process;

– the power law expression [10.2] does apply to the range of the low values of S. The value of statistical parameters depends on the range of flaw strengths.

10.4. Matrix fragmentation: series system model

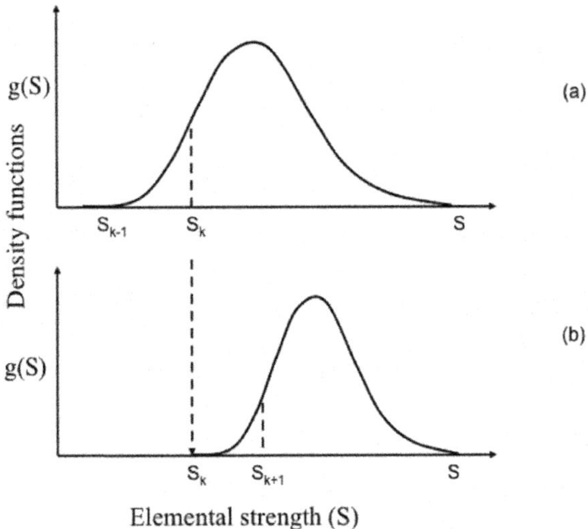

Figure 10.4. *Truncated flaw strength density functions g(S) after formation of a) the (k-1)th crack from flaw having elemental strength S_{k-1}, b) the kth crack from flaw having elemental strength S_k*

The damage process by successive failure during monotonous loading displays the features of a series system, as indicated above. The stressed volume remains unchanged during damage. The elimination of flaws as cracking proceeds is expressed by truncating the flaw density function g(S) [LAM 10]. Thus, after formation of the kth crack generated by a flaw having strength S_k, the relevant density function for the formation of the next crack is $g(S > S_k)$ (Figure 10.4). The probability of failure from defects remaining after the formation of the k crack is:

$$P_t = \frac{P_j - P_k}{1 - P_k} \qquad [10.6]$$

where P_j is the probability of failure before the formation of the crack k. P_j corresponds to the flaw density function $g(S > S_{k-1})$. P_k is the probability for the formation of crack k. P_t is the probability of failure after the formation of the crack k. It corresponds to the flaw density function $g(S > S_k)$.

10.4.1. *Uniform tensile stress state: uniaxial elemental strength approach*

This is the simplest case when the elemental stresses are oriented in the loading direction. A power form flaw strength distribution was selected for the flaw density function, for simplicity, which is not quite satisfactory because it is an extreme values theory-based equation. It may be an acceptable approximation for a small number of cracks and then it will diverge. More accurate strength distributions can be used.

The cumulative distribution function before the formation of the first crack is given by:

$$P(1) = 1 - \exp\left[\frac{V}{V_o}\frac{S^m}{S_o^m}\right] \quad [10.7]$$

The probability of formation of the first crack from a flaw having strength S_1 is denoted as P_1:

$$P_1 = 1 - \exp[-\frac{V}{V_o}\left(\frac{S_1}{S_o}\right)^m] \quad [10.8]$$

From equation [10.6], it comes for the cumulative distribution function after formation of the first crack:

$$P(2) = \frac{P(1)-P_1}{1-P_1} \quad [10.9]$$

$$P(2) = 1 - \exp\left[-\frac{V}{V_o}\frac{S^m - S_1^m}{S_o^m}\right] \quad [10.10]$$

The probability of formation of the second crack is expressed as:

$$P_2 = 1 - \exp\left[-\frac{V}{V_o}\frac{S^m - S_1^m}{S_o^m}\right] \quad [10.11]$$

where S_2 is the strength of the flaw that initiated the second crack.

Similarly for the kth crack, it comes:

$$P(k) = \frac{P(k\text{-}1)\text{-}P_{k\text{-}1}}{1\text{-}P_{k\text{-}1}} \quad [10.12]$$

$$P(k) = 1\text{-}\exp\left[-\frac{V}{V_o S_o^m}(S^m - S_{k\text{-}1}^m)\right] \quad [10.13]$$

$$P_k = 1 - \exp\left[-\frac{V}{V_o S_o^m}(S_k^m - S_{k-1}^m)\right] \quad [10.14]$$

The strengths of those flaws that induced cracks are derived from the expressions of probabilities of crack formation as:

$$S_1^m = -\frac{S_o^m V_o}{V}\text{Ln}(1 - P_1) \quad [10.15]$$

$$S_2^m - S_1^m = -\frac{V_o V_o^m}{V}\text{Ln}(1 - P_2) \quad [10.16]$$

$$S_{k-1}^m - S_{k-2}^m = -\frac{V_o V_o^m}{V}\text{Ln}(1 - P_{k-1}) \quad [10.17]$$

$$S_k^m - S_{k-1}^m = -\frac{V_o S_o^m}{V}\text{Ln}(1 - P_k) \quad [10.18]$$

Summing [10.15]–[10.18] gives:

$$S_k^m = -\frac{V_o S_o^m}{V}[\text{Ln}(1 - P_1)(1 - P_2)(1 - P_{k-1})(1 - P_k)] \quad [10.19]$$

P_1, P_2, ... P_k correspond to the low extremes of the successive distribution functions. Thus, assuming that P_α can be defined such that:

$$(1-P^1)(1-P^2)\ldots(1-P^k) = (1-P_\alpha)^k \qquad [10.20]$$

equation [10.21] reduces to:

$$S_k = S_0\, k^{\frac{1}{m}} \left(\frac{V_0}{V}\right)^{\frac{1}{m}} [-\mathrm{Ln}(1-P_\alpha)\,]^{\frac{1}{m}} \qquad [10.21]$$

P_α may be taken to be 1%, 0.1% or less, depending on the flaw population size. P_α may be also selected randomly, or k-dependent, but it must remain quite small. To some extent, the value of P_α depends on the number of flaws in the distribution. It is assumed that the number of remaining flaws is much larger than 1 so that statistics may apply.

S_o, V and P_α are the constants, thus we find: $S_K \propto k^{1/m}$, and:

$$S_k^* = \frac{S_k}{S_o}\,[-\frac{V_o}{V}\mathrm{Ln}(1-P_1)\,]^{-1/m} = k^{-1/m} \qquad [10.22]$$

Figure 10.5 shows the plots of equation [10.22] for different values of shape parameter m. Note that the damage rate slows down more rapidly as m increases.

10.4.2. *Non-uniform stress state: multiaxial elemental strength approach*

Failure probability is given by general equation [7.1] that was rewritten as:

$$P = 1 - \exp\,[-\frac{V}{V_o} K \left(\frac{\sigma_1}{S_o}\right)^m\,] \qquad [10.23]$$

where K depends on the stress state and on the expression of flaw density function. σ_1 is the principal stress. As discussed in Chapter 4, the underlying flaw density function is defined as a function of the multiaxial elemental strength as: $g_E(S_E) = mS_E^{m-1} S_o^m$.

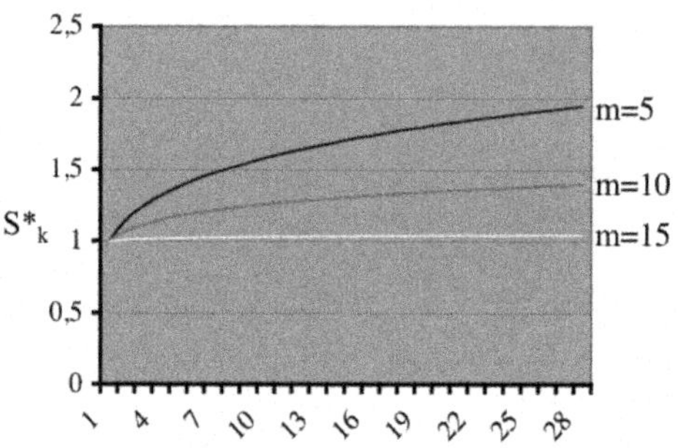

Figure 10.5. *Normalized elemental strengths S_k^* as a function of number of matrix cracks (k) for various values of shape parameter m*

The approach developed in the previous section is valid for equation [10.23]. The probability of formation of the kth crack is similar to equation [10.14], with σ_1 instead of S:

$$P_k = 1 - \exp\left[- \frac{V}{V_o} K \left(\frac{\sigma_1^m - \sigma_{1(k-1)}^m}{S_o^m} \right) \right] \qquad [10.24]$$

where $\sigma_{1(k-1)}$ is the principal stress component required to create the (k-1)th crack. The stresses at formation of the successive cracks are derived from equation [10.24]:

$$\sigma_{1k} = S_0 \ [-\mathrm{Ln}(1-P_1)(1-P_2)(1-P_k)\,]^{\,1/m} \qquad [10.25]$$

10.4.3. *Influence of flaw strength density function*

The initial cumulative distribution function as well as the successive truncated distributions was based on the same power law of flaw strength density function. We may wonder whether this expression represents the whole population of critical defects as well as the defect populations remaining as the number of cracks increases. Various alternate expressions are discussed in Chapter 3 (section 3.3) that can be introduced into equation [4.45] for failure probability determination.

It is then necessary to determine the successive cumulative distribution functions after the formation of cracks. After the formation of the ith crack, the cumulative distribution function is defined as:

$$P(i) = 1 - \exp\left[-\frac{1}{V_o}\int_v dV \int_{S_{i-1}}^{S_i} \frac{1}{2\pi}\int_0^{\frac{\pi}{2}}\int_0^{\pi} g_E(S_E > S_E^{i-1})dS_E \cos\phi d\phi d\psi \right] \quad [10.26]$$

The probability of formation of the ith crack (P_i) is given by equation [10.27] for $S_E = S_E^i$. The successive cumulative distribution functions are derived from equation [10.6] as:

$$P\,(i+1) = \frac{P(i) - P_i}{1 - P_i}$$

With the Weibull flaw strength density function, it comes for the flaw strength density function:

$$g_E(S_E) = m\frac{S_E^{m-1}}{S_o^m}\exp\left[-\left(\frac{S_E}{S_o}\right)^m\right] \quad [10.27]$$

Substituting expression [4.50] for S_E, leads to:

$$g_E(S_E) = m\,\frac{\sigma_1^{m-1} f_v^{m-1}}{S_o^m}\exp\left[-\left(\frac{\sigma_1 f_v}{S_o}\right)^m\right] \quad [10.28]$$

The probability of formation of cracks can be computed using a numerical analysis. The procedure of estimation of statistical parameters can use simulation of the process of successive ruptures, the series of stresses of formation of cracks being compared with experimental results.

10.5. Approach based on Poisson process

In probability theory, a Poisson process is a stochastic process that counts the number of events and the time points at which these events occur in a given time interval. In the spatial Poisson process, the random variables are defined as the counts of the number of events inside each of a number of finite volumes. An event is defined as the occurrence of a defect in a point the position of which is identified by the space variable. It is assumed that defects are distributed in the material following a Poisson process. This approach was applied to the fragmentation of a fiber within a polymer matrix or metal. The distribution of defects is based on the following assumptions [JEU 91, BAX 94]:

1) Points like defects are considered.

2) The number of defects which appear in a volume (V, V + dV) is independent of the number of defects in the initial volume (0, V) (no history).

3) There are no multiple defects.

4) The number of defects is independent of the position of volume.

The number of defects in volume V is a Poisson random variable with parameter θV. θ is a positive real number representing the average number of defects per unit of volume, also referred to as density or intensity. A volume V of material possesses an infinite resistance except at points defined by the Poisson process. The number of defects the resistance of which is lower than the applied stress σ (uniform field) is equal to $\theta(\sigma)V$. The probability of finding N activated defects (N cracks) in the volume V is given by:

$$P(N = n) = \left(\frac{(\theta(\sigma)V)^n}{n!}\right) \exp[-\theta(\sigma)V] \quad [10.29]$$

When fracture occurs after a critical number of cracks have formed from a critical number of defects N_c, failure probability is given by:

$$P(N > N_c) = 1 - \sum_{n=0}^{N_c} \frac{(\theta(\sigma)V)^n}{n!} \exp[-\theta(\sigma)V] \quad [10.30]$$

with $N_c = \theta_c(\sigma)V$. θ_c is the critical density of defects.

For brittle fracture in the absence of damage (no defect is activated in V (N = 0) prior to material fracture), we obtain the expression of failure probability for a uniform stress state:

$$P = 1 - \exp\,[-\,\theta(\sigma)V] \qquad [10.31]$$

The following formulation for the fragmentation of fibers was proposed [HEN 89, HUI 95]. The probability of fracture of a fiber of length L (uniform stress field) is given by the following equation:

$$P(\sigma, L) = 1 - \exp\,[-\,L \int_0^{\sigma} \mu(s)\,ds] \qquad [10.32]$$

where $\mu(\sigma)$ is the intensity of Poisson process.

The intensity of Poisson process depends on the set of flaw strengths (and particularly the limits S_{min} and S_{max}) and on the density of flaws along fiber. Generally, the density of defects in fiber is not known. Several arbitrary forms of $\mu(s)$ or $\theta(\sigma)$ have been proposed. The following expression corresponds to the extreme values of the Weibull density function discussed previously:

$$\mu(s) = \frac{m}{L_o}\frac{S^{m-1}}{\sigma_0^m} \qquad [10.33]$$

Expression [10.33] leads to the classic equation of Weibull, for a fiber of length L [HEN 89]:

$$P\,(\sigma, L) = 1 - \exp\,[-\frac{L}{L_0}\left(\frac{\sigma}{\sigma_0}\right)^m] \qquad [10.34]$$

It was pointed out that equation [10.33] supposes implicitly that the upper limit of the failure stress interval is infinite. Henstenberg considers that in real materials this upper limit is generally much higher than the observed failure stresses, and that this discrepancy introduces a small error. Other limits of this expression are discussed earlier in this chapter. Nevertheless, this expression appears in fragmentation models [HEN 89, HUI 95].

Various expressions of $\theta(\sigma)$ have been proposed [BAX 94]:

– Two-parameter power function (Weibull expression):

$$\theta(\sigma) = \left(\frac{\sigma - \sigma_s}{\sigma_0}\right)^m \qquad [10.35]$$

– Four-parameter power function (bimodal Weibull expression):

$$\theta(\sigma) = \left(\frac{\sigma - \sigma_s}{\sigma_0}\right)^{m_1} + \left(\frac{\sigma - \sigma_s}{\sigma_{02}}\right)^{m_2} \qquad [10.36]$$

– Sigmoid function:

$$\theta(\sigma) = A\left(1 - \exp - \left(\frac{\sigma - \sigma_s}{\sigma_0}\right)^m\right) \qquad [10.37]$$

where A = maximum density

– Duxburry function:

$$\theta(\sigma) = A\ \exp\left(-\left(\frac{\sigma_0}{\sigma}\right)^m\right) \qquad [10.38]$$

Baxevanakis [BAX 93] used the critical density model for the estimation of statistical parameters from results of fragmentation tests on two types of carbon fibers. The best fit to experimental cumulative density function $\theta(\sigma)$ was obtained for the sigmoid function The experimental data did not fit the power law function. However, calculation of the density function required assumptions on the stress state due to fiber debonding and difficulty in estimating the length of the exclusion zone. Results showed that the spatial distribution of defects could be described by the Poisson model far from the saturation limit [WAG 89, WAG 90, BAX 93].

10.6. The Monte Carlo simulation method

The chain-of-segments model has been widely used for Monte Carlo simulation of fragmentation of fibers in a polymer or metal matrix. The distribution of element strengths is constructed from the weakest link failure of independent fibers. Fibers are assumed to follow the Poisson–Weibull model. As a result, fiber elements as well as a single fiber tension tested at arbitrary gauge length follow equation [10.34].

In Monte Carlo simulations, a fiber length L is subdivided into N_e segments, each of length $\frac{L}{N_e}$ << L. Each segment is randomly assigned strength according to the Weibull cumulative probability distribution derived from [10.34]:

$$P\left(\sigma, \frac{L}{N_e}\right) = 1 - \exp\left[-\frac{L}{N_e L_o}\left(\frac{\sigma}{\sigma_o}\right)^m\right] \quad [10.39]$$

A uniform stress σ_f is gradually applied along the entire fiber length, and a break is formed at a segment when the stress on the segment equals the assigned strength of that segment.

In some cases, very large N_e was selected arbitrarily: N_e = 2,000 [OCH 95], 1,000 [OKA 08] or 10^4 [CUR 94]. In other cases, segment length was the final fragment size at saturation, or was defined as twice the slip length at characteristic strength [HEN 89, CUR 91].

Stress on segment may be determined using various models for shear stress [CUR 94, OKA 08, ROU 09], or the slip is zone considered to be an exclusion zone.

This Monte Carlo simulation procedure involves the following two operations with opposite effects:

– first, distribution of the strengths from the low extreme of the flaw strength density function; which tends to underestimate flaw strengths of the actual population. The equation [10.39] corresponds to failures from the most severe flaws, whereas less severe flaws that become critical under high stresses during fragmentation are discarded;

– second, scaling up of the strengths when using equation [10.39]; which tends to overestimate the strengths. Although the stress at the first failure should correspond to the initial volume V, it is predicted for much smaller volume V_e. The same comment applies as long as fragments have volume larger than V_e.

The predicted strengths depend on the balance between both operations. Thus, they depend on the flaw strength density function, and the number of elements.

10.7. The fragment dichotomy-based model (parallel system)

The fragment dichotomy model considers probable successive fragment sizes that decrease with increasing load. It reproduces the process of fragment generation: a fragment results from the failure of a parent fragment from the most severe flaw [LAM 09, GUI 96, LIS 97b, PAI 05]. The cumulative distribution function for a population of fragments having volume V is given by equation [7.1]. From equation [7.1], it comes for the fracture stress of fragment i:

$$\sigma_{ref}^{i} = \lambda_0 \left[\frac{V_0}{2l_i S_i K} Ln\left(\frac{1}{1-P_i}\right)^m \right] \qquad [10.40]$$

where $2l_i$ is the length of fragment i, S_i is the cross-sectional area and P_i is the failure probability for this segment.

The fragmentation process is characterized by a set of stresses σ_{ref}^{i}. Calculation of the series of stresses σ_{ref}^{i} requires the successive fragment volumes V_i and their failure probabilities P_i.

In order to make a comprehensive presentation of the approach, let us consider uniaxial tensile loading conditions and specimens with a uniform cross-section (Figure 10.1(a) or (b)). As a result, the fragments have simple shapes such as cylinders (during the fragmentation of fibers in polymer matrix composites (PMCs)), hollow cylinders (during matrix fragmentation in ceramic matrix composites (CMCs) reinforced by a single fiber) or cubes or parallelepipeds in multilayers. Thus, $V_i = S_i L_i$, where S_i is the fragment sectional area and L_i is the fragment length. The length of a fragment depends on the location of the critical flaw in the parent fragment. Since flaws have a random distribution in location and size, the length of successive fragments is a statistical variable. It can be derived from the *spatial* distribution of flaw strengths within the fragment using the following equation:

$$P(L_i) = \int_{L_{i-1}} h(x)dx \qquad [10.41]$$

where $P(L_i)$ is the probability that fragment length is L_i, h(x) describes the spatial density of flaw strengths along the fragments and x is the abscissa along fragment axis. L_i is the length of the fragment generated from the failure of a parent fragment with length L_{i-1}. Fragment length is given by the reciprocal function:

$$L_i = [P(L_i)]^{-1} \qquad [10.42]$$

Rupture occurs in that fragment having fracture stress smaller than the stress operating on the fragment:

$$\sigma \geq \sigma_{ref}^{i} (P_i) \qquad [10.43]$$

The average behavior is described by the series of average values of σ_{ref}^{i} given either by $P_i = 0.5$, or by

$$\bar{\sigma}_{ref}^{i} = \frac{\lambda_0 \Gamma(1+1/m)}{K^{1/m} V_i^{1/m}} \qquad [10.44]$$

10.7.1. *Fragmentation of fibers (uniform stress state)*

Analysis of the fragmentation of a fiber in a polymer matrix proposed by Wagner [WAG 90] is based on the same interpretation of phenomenon. The authors consider that the stress field is uniform and that fragments failure follows the Weibull law. Mean stresses of rupture of the fragments are then given by:

$$\bar{\sigma}_{ref} = \sigma_0 \bar{L}^{(-1/m)} \Gamma\left(1+\frac{1}{m}\right) \qquad [10.45]$$

where $\bar{L}$ is the average length of fragments. Taking logarithmic of expression [10.45], it comes:

$$\text{Ln } \bar{L} = -\text{ m Ln } \bar{\sigma} + \text{m Ln } \sigma_o + \text{Ln } [\Gamma(1 + \frac{1}{m})] \qquad [10.46]$$

Nevertheless, this equation did not fit the results of fragmentation test on T300 fiber in an epoxy matrix [WAG 90]. Wagner drew the conclusion that the density of defects along fibers does not follow a Weibull law. However, this statement must be tempered because the stress field was considered being uniform. Taking account of the stress gradients induced by fiber debonding could improve the results.

10.7.2. *Fragmentation of the matrix in unidirectionally reinforced ceramic matrix composites*

As an example, we consider composites reinforced with parallel continuous fibers, and subjected to a tensile load parallel to fiber axis. When a crack is created in the matrix, it is perpendicular to the load direction, and it propagates through the matrix, leading to two new fragments with cross-section S_m. The stress state is affected by fiber/matrix debonding (Figure 10.6). Stresses are given by:

– in the vicinity of interfacial matrix cracks: $u + l_o \leq x < l_d$

$$\sigma_m(x) = \sigma_m \frac{x-u-l_o}{l_d-l_o} \qquad [10.47]$$

$$\sigma_f(x) = \sigma_f \left(1 + a - a\frac{x-u-l_o}{l_d-l_o}\right) \qquad [10.48]$$

– in the rest of fragment: $u + l_d < x < 2\,l_i - (u + l_d)$

$$\sigma_m = \frac{\sigma}{V_m}\frac{a}{1+a} + \sigma_m^{th} \qquad [10.49]$$

$$\sigma_f = \frac{\sigma}{V_f}\frac{1}{1+a} + \sigma_f^{th} \qquad [10.50]$$

where σ_m and σ_f are, respectively, the stresses operating on the fiber and the matrix, u, l_d and l_o are defined in Figure 10.6, $2l_i$ is the length of fragment i, σ is the remote stress applied to the specimen, V_f and V_m are the volume fractions of fiber and matrix, respectively, $a = \frac{E_m V_m}{E_f V_f}$ is the load sharing parameter, E_f and E_m are the fiber and matrix Young's moduli, σ_m^{th} and σ_f^{th} are the residual stresses, respectively, in the matrix and the fiber.

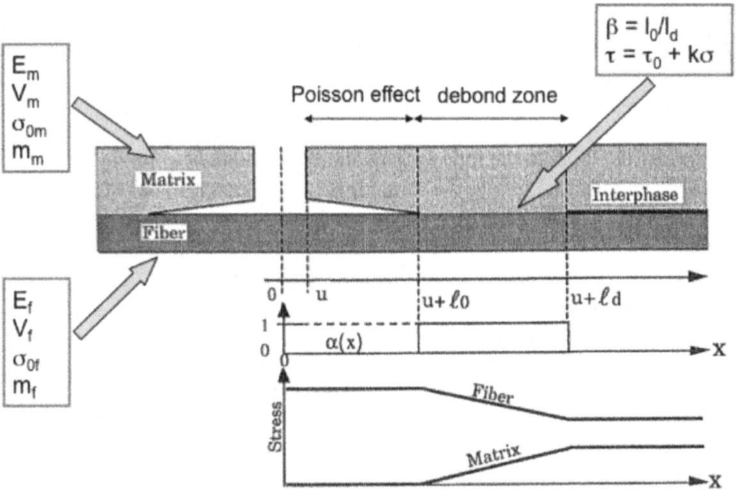

Figure 10.6. *Effect of matrix crack and associated fiber debonding on the stress state operating on the matrix and on the fiber. Also indicated are the matrix and fiber pertinent characteristics*

For the sake of simplicity, the model of fragmentation is based on the Weibull expression of failure probability (equation [5.1]). Equations of the multiaxial elemental strength model can also be used (equation [5.5]). For the above expressions of stress (equations [10.47] and [10.49]), it comes for the failure probability of fragment i:

$$P_{Mi} = 1 - \exp\left[-2S_m \left(\frac{\sigma_m}{\sigma_{om}}\right)^m l_{ei}\right] \qquad [10.51]$$

$$\text{with } l_{ei} = l_i \left(1 - \frac{l_{di}}{l_i}\frac{m+\beta}{m+1}\right) \text{ when } l_i > l_{di} \qquad [10.52]$$

$$l_{ei} = \frac{(l_i - \beta l_{di})^{m+1}}{(m+1) l_{di}^{\,m} (1-\beta)^m} \text{ when } l_i \leq l_{di} \qquad [10.53]$$

$$\beta = \frac{l_o}{l_d}$$

where l_0 is the length of fiber without contact to the matrix as a result of Poisson fiber lateral contraction.

The strength of fragment i having length $2l_i$ (or l_i at the ends of specimen) is thus:

$$\sigma_{Mi}(l_i, P_{Mi}) = \sigma_M^R(P)\left(\frac{V_{OM}}{V_i}\right)^{1/m}\left(\frac{l_{ei}}{l_i}\right)^{-1/m} \qquad [10.54]$$

where V_{OM} is the initial volume of matrix (in the absence of cracks), V_i is the volume of fragment, σ_M^R is the reference matrix strength at probability P and for volume V_{OM}, under a uniform stress state:

$$P = 1 - \exp\left[-V_{OM}\left(\frac{\sigma_m^R}{\sigma_{om}}\right)^m\right] \qquad [10.55]$$

The statistical parameters m and σ_0 can be estimated from the distribution of strengths measured on a set of specimens tested separately.

10.7.2.1. *Approximations and simplifications*

In [LIS 97b], it was assumed that fragments fractured between debond crack tips. A uniform spatial distribution of fracture-inducing flaws was considered:

$$h(x) = \text{constant} = \frac{1}{2l_i - 2l_d} \qquad [10.56]$$

The probability of location of a fracture-inducing flaw in fragment is thus given by:

$$P(x) = \frac{x - l_d}{2(l_i - l_d)} = X \qquad [10.57]$$

where x is the distance to fragment edge defined by matrix crack. X is a random number ($0 < X < 1$).

In [PAI 05], the portion of matrix adjacent to debond cracks was not discarded. It was assumed that the density of probability of location was commensurate to the stress gradient.

Simpler analytical equations can be derived from the above general equations, for the following particular values: either $P_{Mi} = 0.5$, or $X = 0.25$ (cracks in the middle of fragments), or $l_i > l_{di}$. $P_{Mi} = 0.5$ and $X = 0.25$ have been shown to provide satisfactory estimates of stresses and strains. $l_i > l_{di}$ is a restriction condition, since small fragments ($2\ l_i < 2\ l_{di}$) are discarded.

For the above three particular values of P_{Mi}, X and l_i, equation [10.54] reduces to, as $\frac{V_{oM}}{V_i} = n + 1$ at step i:

$$\bar{\sigma}_M^{\,i} = \bar{\sigma}_M^{\,R}\, L^{1/m_m} \left(\frac{L}{n+1} - 2l_d \frac{m+\beta}{m+1} \right)^{-1/m_m} \qquad [10.58]$$

where $\bar{\sigma}_M^{i}$ and $\bar{\sigma}_M^{R}$ are the strengths for $P_{Mi} = 0.5$, L is the gauge length, n is the number of cracks and i indicates the load step number. Note that here $n = 2^i - 1$.

The number of cracks at saturation is obtained when the term between brackets is negative.

$$n_S = \frac{L}{2l_d}\frac{m+1}{m+\beta} - 1 \qquad [10.59]$$

$$n_S \approx \frac{L}{2l_d}$$

The stress at matrix cracking saturation is then derived from equation [10.54], assuming that saturation is reached when $l_i = l_{di}$:

$$\bar{\sigma}_M^{s} = \bar{\sigma}_M^{R} (n+1)^{1/m_m} \left(\frac{1+m_m}{1-\beta}\right)^{-1/m_m} \qquad [10.60]$$

10.8. Evaluation of models: comparison to experimental data

Figure 10.7 shows the nonlinear stress–strain curve dictated by matrix fragmentation for SiC/SiC composite specimens that was predicted using the fragment dichotomy model ($P_{Mi} = 0.5$) [LAM 09]. The prediction compares fairly well with the experiments. Figure 10.8 shows that taking random numbers does not improve results. This confirms that good approximation of behavior is obtained for $P_{Mi} = 0.5$ [LIS 97b, PAI 05]. It can also be noticed from Figure 10.8 that predictions are satisfactory with $P_{Mi} = 0.5$ and $X = 0.25$.

Figure 10.9 shows a logarithmic plot of fragment strength-size data measured using scanning electron microscopy on several SiC/SiC single fiber reinforced composite specimens. The strengths obtained for identical fragment sizes display a statistical distribution. Thus, strengths increase, as fragments are smaller. The best fit line follows the linearized equation of size effect given in Chapter 6 for Weibull modulus m = 7.5:

$$\text{Ln}\,\sigma_{Mi} = -\frac{1}{7.5}\,\text{Ln}\,V_i + \text{constant} \qquad [10.61]$$

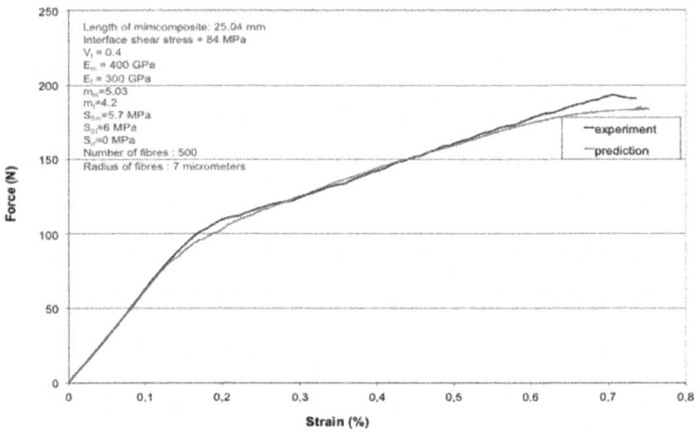

Figure 10.7. *Tensile behavior for a unidirectionnally reinforced composite SiC/SiC composite: comparison of experimental result and prediction using the fragment dichotomy-based model*

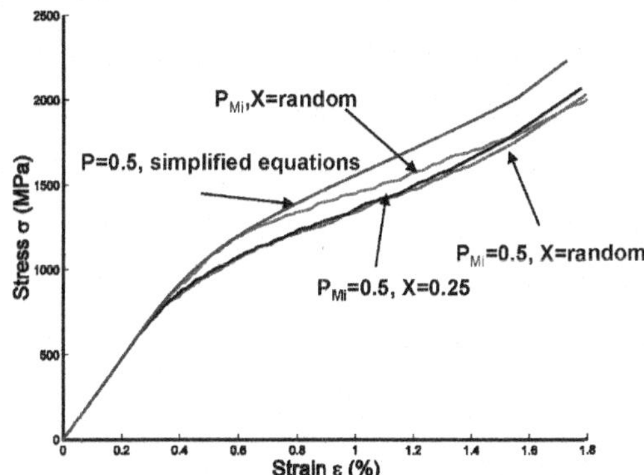

Figure 10.8. *Stress-strain curves for SiC/SiC single fiber reinforced composite predicted using the fragment dichotomy-based model for various values of matrix fragment failure probability (P_{Mi}) and crack location probability (X), and using the simplified equation of fiber failure probability*

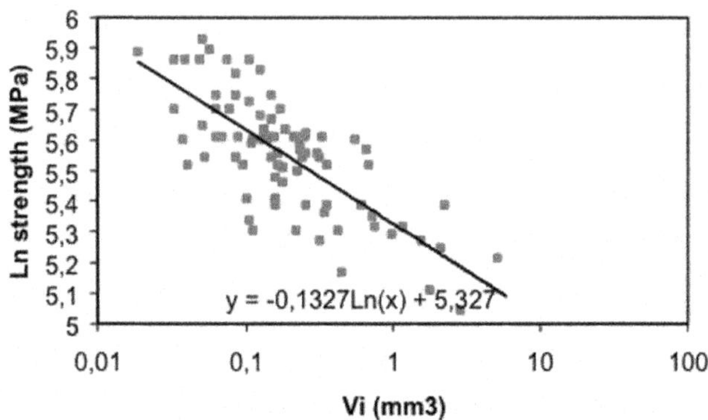

Figure 10.9. *Logarithmic plot of fragment strengths versus volumes during fragmentation of the SiC matrix of a unidirectionnally reinforced composite specimen. The solid line represents the best fit curve*

This Weibull modulus estimate compares fairly well with values extracted from the statistical distributions of matrix strengths at onset of cracking determined on two sets of SiC/SiC minicomposites [LIS 97b]: 5.1–6.2 (standard deviation: 1.0 and 1.9). These results demonstrate that failure of fragments is described satisfactorily by an extreme-values theory-based model. This implies that the failure of a fragment is dictated by the most severe among the flaws present within the fragment, and that the flaw density function is independent of fragment size. This latter hypothesis is implicit in the Weibull model. Note from Figure 10.9 that the trend is observed down to fragment lengths as small as 60 μm.

Figure 10.10 compares the distribution of fragmentation stresses obtained experimentally to predicted fragments or element strengths. The experimental distribution was obtained by plotting normalized cumulative acoustic emission data that is the cumulative number of counts divided by the total cumulative number of counts at the final event, versus the stress on the matrix σ_m. For predictions using equations [10.21] and [10.58], failure probability was given by n/n_s:

(n = number of cracks, n_s = number at saturation) with n_s derived from equation [10.59].

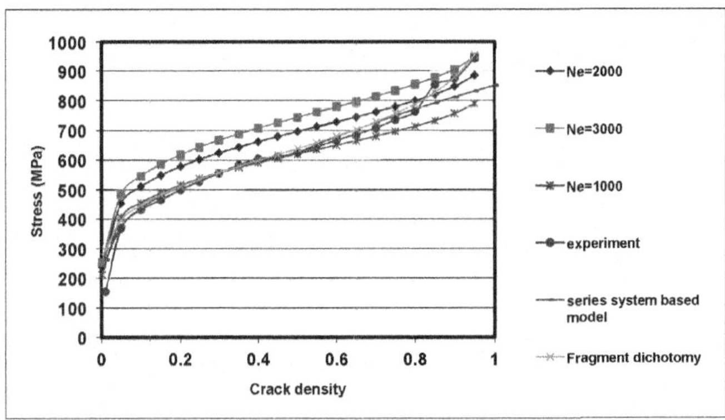

Figure 10.10. *Distributions of fragment or element strengths for the SiC matrix in SiC/SiC unidirectionnally reinforced composite. Comparison of experiment with predictions using the Monte Carlo simulation method, the series system-based model and the fragment dichotomy-based model*

Figure 10.10 shows that the experimental distribution agrees with predictions by the fragment dichotomy model, as well as the truncated flaw population-based model (equation [10.21]) for m ≈ 5 and $S_0 \approx 12$. These m and S_0 estimates were not significantly affected by P_α slight increase that was aimed at accounting for decrease in the number of flaws in the matrix as fragmentation proceeds.

Such a good agreement with the truncated flaw population-based model may appear to be surprising since the power form may be expected not to be suitable to describe the whole distribution of flaws. It seems that the power form was sufficient here, and that values of parameters m and S_0 can be found so that the strength interval is in agreement with the number of flaws which were activated during the fragmentation process.

By contrast, the number of elements N_e affects predictions by the Monte Carlo simulation method. When $N_e > n_s$, the element strengths were overestimated. This reflects strength scaling up by equation

[10.39], since the volume of elements was much smaller than that of fragments. When N_e (= 1000) < n_s (= 1246), the predicted strengths agreed better in the range of low stresses (< 625 MPa), although they slightly overestimated the low extreme, which suggests that the underestimation of strengths by the power law was balanced by the scaling up.

Available literature results on frequentation of fibers show that predictions using the Monte Carlo simulation method are very sensitive to the values of statistical parameters introduced in equation [10.39]. Agreement was not obtained with those parameters derived from the failure of independent fibers with gauge length L [OKA 08, ROU 09]. Other experimental results showed that predictions are satisfactory only for low stresses quite far from the saturation of fragmentation [BAX 93, OKA 99].

10.9. Ultimate failure of unidirectionnally reinforced composite (Weibull model, uniform tension) in the presence of matrix damage

Ultimate failure is derived from the survival probabilities of fiber volume elements subjected to the stress state defined by equations [10.48] and [10.50] [LIS 97b]. According to the weakest link principle, the failure probability is obtained from the product of no failure probabilities of fiber volume elements:

$$P = 1 - \underset{k}{\pi}\,(1 - P_i)\,\underset{t}{\pi}\,(1 - P_j) \qquad [10.62]$$

with $P_w = 1 - \exp\left[-\left(\frac{\sigma_f}{\sigma_{of}}\right)^{m_f} S_f\, l_{e_w}^{f}\right]$

where w = i for those elements where $l_{di} < l_i$, and w = j for those where $l_{di} > l_i$. S_f is the fiber cross-sectional area, σ_f is the stress operating on fiber (equation [10.50]), l_{ew}^{f} is the equivalent length.

Failure probability of the fiber is:

$$P_f = 1 - \exp\left[-\left(\frac{\sigma_f}{\sigma_{of}}\right)^{m_f} S_f \left(\sum_k l_{ei}^{f} + \sum_t l_{ej}^{f}\right)\right] \qquad [10.63]$$

where l_{ei}^f and l_{ej}^f are the equivalent lengths, respectively, for the k elements where $l_{di} \leq l_i$, and the t are elements where $l_{di} > l_i$.

An equation of fiber strength can be derived from equation [10.63], in terms of reference fiber strength σ_f^R (P_f), using the same method as above for matrix fragments:

$$\sigma_f(P_f) = \sigma_f^R(P_f)\left[\sum_k \frac{l_{ei}^f}{L} + \sum_t \frac{l_{ei}^f}{L}\right]^{1/m} \qquad [10.64]$$

Then, for $l_i \geq l_{di}$, the following equation of fiber failure probability can be derived from equation [10.63], in the presence of n cracks in the matrix:

$$P_f = 1 - \exp\left[-\left(\frac{\sigma_f}{\sigma_{of}}\right)^{m_f} V_F\left(1+\frac{2n\bar{l}_{di}}{L}A\right)\right] \qquad [10.65]$$

where $\bar{l}_{di}$ is the average debond length, V_F is the fiber volume.

$$A = \left(\frac{u}{l_d}+\beta\right)(1+a)^{m_f} + \left[\frac{1-\beta}{m+1}\left(\frac{(1+a)^{m_f+1}-1}{a}\right)-1\right] \qquad [10.66]$$

A is an increasing function of load sharing parameter a. After saturation of matrix cracking ($l_i = l_{di}$), equation [10.65] leads to the following equation of stress on fiber:

$$\sigma_f = \sigma_{of}\left[\frac{Ln\dfrac{1}{1-P_f}}{V_F(1+A)}\right]^{1/m_f} \qquad [10.67]$$

10.10. Application to composites: unified model

10.10.1. *Uniform tension, unidirectional composites and the Weibull model*

Failure probability equations provide criteria for the occurrence of fragmentation whatever the matrix and the fibers. They lead to a unified model of multiple cracking for polymer or ceramic matrix composites.

When the load sharing parameter a = $\frac{E_m V_m}{E_f V_m}$ is small (in polymer matrix composites), σ_m is negligible to σ_f (equations [10.49] and [10.50]). Probabilities of the first failures for the fiber and the matrix are given by:

$$P_M \approx 0$$

$$P_f = 1 - exp\left[-V_F\left(\frac{\sigma_f}{\sigma_{of}}\right)^{m_f}\right] \qquad [10.68]$$

$P_M = 0$ and $P_f > 0$ indicate that cracking occurs first in the fiber only.

When the load sharing parameter a is initially significantly large ($E_m > E_f$ in ceramic matrix composites), σ_f is now negligible (equation [10.50]). Probabilities of first failure for the fiber and the matrix are given by:

$$\begin{cases} P_M = 1 - exp\left[-V_M\left(\dfrac{\sigma_m}{\sigma_{oM}}\right)^{m_m}\right] \\ P_f \approx 0 \end{cases} \qquad [10.69]$$

Equations [10.69] indicate that cracking occurs first in the matrix. Then, P_f increases during matrix cracking (equation [10.65]). After saturation of matrix cracking, fiber strength is given by equation

[10.67]. When the contribution of matrix to load sharing approaches 0, A $\rightarrow$ 0 and equation [10.67] tends to:

$$\sigma_f = \sigma_{of}\left[\frac{1}{V_F}Ln(\frac{1}{1-P_f})\right]^{1/m} \qquad [10.70]$$

From the comparison of equations [10.65], [10.67] and [10.70] at a given value of P_f, it appears that σ_f increases with decreasing a. Furthermore, equation [10.70] of maximum stress is equivalent to equation [10.68] that defines the failure probability of a fiber subjected to a uniform tension, which represents the worst loading condition, as discussed previously in Chapter 6. It can be deduced that fibers should not fail prior to saturation of matrix cracking, unless extra loads or localized phenomena make fiber failure probability to increase.

The same conclusion is reached from the severity factor $K = V_E/V$ defined in Chapter 6. $K < 1$ is the presence of stress gradients. Consequently, the stress required for failure of fiber prior to matrix cracking saturation is greater than the strength of fiber under maximum stress operating uniformly. Thus, the fiber does not fracture while $K = 1$ is not reached at matrix cracking saturation. This conclusion is supported by experimental results [LIS 97a, LAM 01].

The same analysis can be applied to brittle polymer matrix. It can be shown that P_M increases as fiber fragmentation proceeds. However, with some PMCs, other phenomena such as changes in elastic properties, plasticity, must be taken into account. The behaviour of other material combinations for which the load sharing parameter (a) takes intermediate values can be investigated by computing P_M and P_f. Furthermore, it is worth mentioning that σ_o and m may influence the trend at intermediate values of a. Each situation depends on constituent properties. It can be computed using the above general equations, taking into account the stress state in the constituents and properties of constituents.

10.10.2. *General approach, the multiaxial elemental strength model*

Since failures are caused by defects, the flaw strength density functions pertinent to composite constituents control damage and the ultimate fracture. The behavior of fiber reinforced composite materials results from concurrent damage phenomena in the constituents, i.e. the matrix and the fibers in unidirectionally reinforced composites. Additional entities such as plies may be involved in multidirectionally reinforced composites. The damage phenomena follow one another in some ceramic matrix composites such as SiC/SiC (as shown in the previous section) and C/SiC (silicon carbide matrix reinforced by carbon fiber). SiC/SiC with two-dimensional woven reinforcement contains additional entities at various length scales, which experience successive cracking modes: the interply matrix, the intraply matrix in the transverse bundles, the intraply matrix in the longitudinal tows (parallel to the loading direction), and finally, the fracture of fibers and tows (Figure 10.1) [LAM 01].

10.10.2.1. *Application to the behaviour of ceramic matrix composites*

Figure 10.11 represents the flaw strength density functions for the matrix and the fibers: respectively, $g_m(S)$ and $g_f(S)$. Additional flaw strength density functions can be introduced to account for the presence of more families of cracks. In Figure 10.11(a), $g_m(S)$ and $g_f(S)$ are distinct in order to illustrate the chronological activation of defects which leads to matrix damage and then to fiber tow fracture. The elemental strengths of the matrix flaws are weaker than those of the fibers. They illustrate the actual situation encountered for the SiC matrix composites. Figure 10.11(b) represents an alternative possible order of elemental strengths. The effect of relative position of flaw strength density functions is shown accurately by the failure probability, which takes into account the effects of size and stress state. The presentation of Figure 10.11 shows the significance of respective flaw density functions for the mechanical behaviour of fiber reinforced ceramic matrix composites. It implies a uniform stress state.

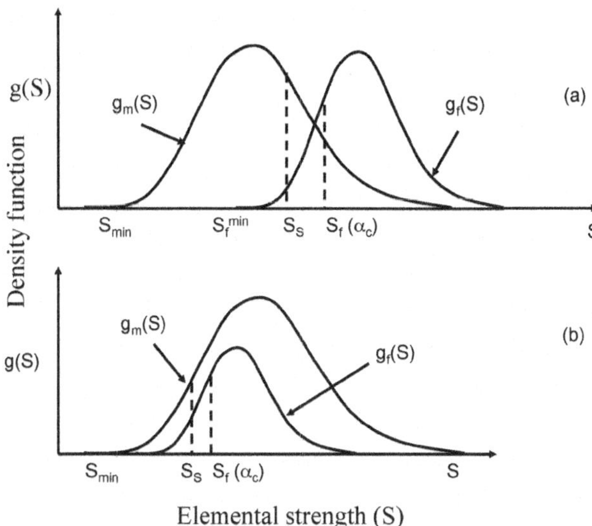

Figure 10.11. *Flaw strength density functions for the matrix ($g_m(S)$) and for the fibers ($g_f(S)$) in a composite material: a) The matrix flaws have lower elemental strengths; b) Both matrix and fiber flaws have comparable severity, S_S is the elemental strength at matrix cracking saturation, $S_f(\alpha_c)$ is the elemental strength in the critical filament at tow failure*

The individual fracture of fibers is dictated by the most severe defects, characterized by elemental strengths at the low extreme of the flaw strength density function $g_f(S)$. Ultimate fracture of tow takes place when a critical number of fibers are broken. The corresponding values of the elemental strengths are then included in the interval [S_f^{min}, $S_f(\alpha_c)$], where α_c is the fraction of fibers broken individually.

10.10.2.2. *Probability of formation of cracks*

The probability of formation of a crack in the matrix is given by equation [10.26], the stress field to be taken into account is that on the matrix. The simplified equations can also be used when appropriate. The probability of failure of a fiber is given by equation [7.1] for the stress state on fibers. The stress field evolves as cracks form in the matrix and as fibers fail as a result of load sharing. The approach discussed in section 10.4 allows us to determine the series of stresses at crack formation.

The computation of failure probability requires expressions for the flaw strength density functions. Data on fiber/matrix interactions and the critical number of individually broken fibers are also required. The method can be repeated with many more flaw strength density functions.

10.10.2.3. *Overall probability of composite damage*

The overall probability of damage of composite is the probability of formation of cracks either in the matrix or in fibers (fibers break). These failures are independent and result from concurrent flaw populations, when considering the volume of composite specimen, as done for a series system. According to the weakest link principle, the failure probability is obtained from the product of non-failure probabilities of fibers and matrix:

$$P = 1 - (1 - P_m)(1 - P_f) \qquad [10.71]$$

where P_m is the probability of formation of a crack in the matrix and P_f is the probability of fracture of a fiber. The equation of the global probability of damage allows handling of all the situations, in particular when the flaw strength density functions are multiple because of the presence of several populations, and when they overlap (intervals of comparable elementary constraints) or are placed in an order different from that of Figure 10.11. However, the case where the elemental strengths of fiber flaws would be weaker than those of the matrix is not realistic for CMCs because by definition the reinforcement is more resistant than the matrix.

10.11. Conclusion

It is worth emphasizing that the elemental strength-based approaches can be extended to modelling the damage mode of composite materials, which result from multiple cracking induced by flaw populations. The damage process can be modelled either as the brittle fracture of successive fragments (equivalent to a parallel system of elements), or as the sequential formation of cracks in the total damaged volume of material (equivalent to a series system). In the latter case, the flaw strength density function must be truncated as flaws are eliminated due to the formation of cracks, and the flaw strength density function must correspond to the whole population of flaws that will initiate cracks. This population is broader than the population of the most severe flaws that is considered in the description of brittle fracture. The pertinent flaw strength distribution function cannot be estimated from the results of fracture tests, but instead from multiple cracking or fragmentation tests.

The fragmentation of a fiber in a polymer matrix may be regarded as the occurrence of point defects according to a Poisson process. Such an approach provides an expression of failure probability as a function of flaw critical density. The validity of this model is questioned by a few authors, and it is restricted to low stresses with respect to the stress at saturation.

According to the brittle fracture scheme, fragmentation of ceramic brittle matrix in ceramic composites is modelled as a multiple test on fragments characterized by a unique flaw strength density function. The validity of the fragment dichotomy model has been demonstrated on SiC matrix composites reinforced with either SiC or carbon fibers. The fragment dichotomy model has been established for uniaxial stress state and power law flaw strength density function. The multiaxial elemental strength model can be introduced to simulate fragmentation under multiaxial load.

However, Wagner did not correlate satisfactorily the results of fragmentation test on T300 fiber in an epoxy matrix to the Weibull distribution of rupture stresses.

The behavior of composite materials containing at least brittle fibers results from at least two main concurrent damage phenomena in the matrix and the fibers. It can be described and modelled using the respective flaw strength density functions and the associated failure probabilities. Depending on the respective properties of matrix and fibers and flaw strength density functions, damage affects first the matrix or the fibers. In ceramic matrix composites, matrix damage and fiber failure are sequential. Statistical-probabilistic approaches provide a unified theory of damage for composite materials.

Bibliography

[AMA 88] AMAR E., GAUTHIER F., LAMON J., "Reliability analysis of a Si_3N_4 ceramic piston for automotive engines", in TENNERY V.J. (ed.), *Proceedings of the 3rd International Symposium Ceramic Materials and Components for Engines*, American Ceramic Society Inc., Las Vegas, USA, pp. 1334–1346, 27–30 November 1988.

[AND 52] ANDERSON T.W., DARLING D.A., "Asymtotic theory on certain goodness of fit criteria based on stochastic process", *Annals of Mathematical Statistics*, vol. 23, pp. 193–212, 1952.

[ARG 59] ARGON A.S., "Surface cracks on glass", *Proceedings of the Royal Society*, vol. 250, pp. 482–492, 1959.

[ARG 66] ARGON A.S., MCCLINTOCK F.A., *Mechanical Behavior of Materials*, Addison-Wesley, Reading, MA, 1966.

[ASH 86] ASHBY M.F., JONES D.R.H., *Engineering Materials 2*, Pergamon Press, Great Britain, 1986.

[BAD 87] BADER M.G., MITH R.L.S., PITKETHLY M.J., "Probabilistic model for hybrid composites", in MATTHEWS F.L., BUSKELL N.C.R., HODGKINSON J.M. (eds), *6th International Conference on Composites Materials*, Elsevier Applied Science, London and New York, 1987.

[BAR 75] BARNETT V., "Probability plotting methods and order statistics", *Journal of the Royal Statistical Society*, vol. 24, pp. 95–108, 1975.

[BAT 74] BATDORF S.B., Fracture statistics of brittle materials with intergranular cracks; Rep. No. SAMSO-TR-74-210, The Aerospace Corp., El Segundo, CA, October 1974.

[BAT 74a] BATDORF S.B., CROSE J.G., "A statistical theory for the fracture of brittle structures subjected to nonuniform polyaxial stresses", *Journal of Applied Mechanics*, vol. 41, pp. 459–464, 1974.

[BAT 74b] BATDORF S.B., "Weibull statistics for polyaxial stress states", *Journal of the American Ceramic Society*, vol. 57, no. 1, pp. 44–45, 1974.

[BAT 77] BATDORF S.B., "Some approximate treatments of fracture statistics for polyaxial tension", *International of Fracture*, vol. 13, pp. 5–11, 1977.

[BAT 78] BATDORF S.B., HEINRISCH H.L., "Weakest link theory reformulated for arbitrary fracture criterion", *Journal of the American Ceramic Society*, vol. 61, nos. 7–8, pp. 355–358, 1978.

[BAX 93] BAXEVANAKIS C., JEULIN D., VALENTIN D., "Fracture statistics of single-fiber composite specimens", *Composite Science and Technology*, vol. 48, pp. 47–56, 1993.

[BAX 94] BAXEVANAKIS C., Comportement statistique à rupture des composites stratifiés (statistical failure of laminated composites), PhD Thesis, Ecole Nationale Supérieure des Mines de Paris, Paris, France, December 1994.

[BER 84] BERGMAN B., "On the estimation of the Weibull modulus", *Journal of Materials Science Letters*, vol. 3, pp. 689–692, 1984.

[BOR 02] BORNEMISZA T., JONES A., "Ceramic gas turbine development at Hamilton Sundstrand power systems", in VAN ROODE M., FERBER M.K., RICHERSON D.W. (eds), *Ceramic Gas Turbine Design and test Experience, Progress in Ceramic Gas Turbine Development*, ASME Press, New York, 2002.

[BOU 82] BOUSSUGE M., BROUSSAUD D., LAMON J., "Delayed fracture of reaction-bonded Si_3N_4 during tensile fatigue experiments", *Proceedings of the British Ceramic Society*, no. 32, pp. 205–212, 1982.

[BOU 83] BOULEAU N., *Splendeurs et misères des lois de valeurs extrêmes – Cinq conférences sur l'indécidabilité*, Presses de l'ENPC, 1983.

[BOU 86] BOUSSUGE M., LAMON J., "Effects of static and cyclic fatigue at high temperature upon reaction-bonded silicon nitride", *Journal de Physique*, no. 2, vol. 47, pp. 557–562, 1986.

[CAL 04] CALARD V., LAMON J., "Failure of fibres bundles", *Composites Science and Technology*, vol. 64, pp. 701–710, 2004.

[CAL 96] CALARD V., LAMON J., "Simulation de la fragmentation de la matrice dans les composites céramiques unidirectionnels", in BAPTISTE D., VAUTRIN A. (eds), *Compte Rendus des Dixièmes Journées Nationales sur les Composites*, vol. 3, AMAC, 1996.

[CUR 91] CURTIN W.A., "Exact theory of fiber fragmentation in a single-filament composite", *Journal of Materials Science*, vol. 26, pp. 5239–5253, 1991.

[CUR 94] CURTIN W.A., NETRAVALI A.N., PARK J.M., "Strength distribution of Carborundum polycrystalline SiC fibers as derived from the single-fiber-composite test", *Journal of Materials Science*, vol. 29, pp. 4718–4728, 1994.

[DAG 86] D'AGOSTINO R.B., STEPHENS M.A., *Goodness of Fit Techniques*, Marcel Dekker, New York, 1986.

[DAV 04] DAVIES I.J., "Best estimate of Weibull modulus obtained using linear least squares analysis: an improved empirical correction factor", *Journal of Materials Science*, vol. 39, no. 4, pp. 1441–1444, 2004.

[DE 77a] DE JAYATILAKA S.A., TRUSTRUM K., "Statistical approach to brittle fracture", *Journal of Materials Science,* vol. 10, pp. 1426–1430, 1977.

[DE 77b] DE JAYATILAKA S.A., TRUSTRUM K., "Application of a statistical method to brittle fracture in biaxial loading systems", *Journal of Materials Science*, vol. 12, pp. 2043–2048, 1977.

[DOR 92] DORTMANS L., DE WITH G., "Weakest-link failure prediction for ceramics. IV: Application of mixed mode fracture criteria for multiaxial loading", *Journal of the European Ceramic Society*, vol. 10, pp. 109–114, 1992.

[DUF 03] DUFFY S.F., JANOSIK L.L., WERESZCZAK A.A. *et al.*, "Life prediction of structural components", in VAN ROODE M., FERBER M.K., RICHERSON D.W. (eds), *Ceramic Gas Turbine Component Development and Characterization*, ASME Press, New York, 2003.

[ECK 89] ECKEL A.J., BRADT R.C., "Concurrent flaw populations in SiC", *J. Am. Ceram. Soc.*, vol. 72, p. 455, 1989.

[EPS 48] EPSTEIN B., "Application of the theory of extreme values in fracture problems", *American Statistical Association Journal*, pp. 403–412, 1948.

[ESH 57] ESHELBY J.D., "The determination of the elastic field of an ellipsoidal inclusion and related problems", *Proceedings of the Royal Society London*, vol. 241, pp. 376–396, 1957.

[ESH 59] ESHELBY J.D., "The elastic field outside an ellipsoidal inclusion", *Proceedings of the Royal Society London*, vol. 252, pp. 561–569, 1959.

[ESH 61] ESHELBY J.D., "Elastic inclusion and inhomogeneities", in SNEDDON I.N., HILL R. (eds), *Progress in Solid Mechanics,* 2nd ed., North-Holland, Amsterdam, 1961.

[EVA 76] EVANS A.G., LANGDON T.G., "Structural ceramics", *Progress in Materials Science*, vol. 21, pp. 320–331, 1976.

[EVA 78] EVANS A.G., "A general approach for the statistical analysis of multiaxial fracture", *Journal of the American Ceramic Society*, vol. 61, no. 7–8, pp. 281–376, 1978.

[EVA 80] EVANS A.G., MEYER M.E., FERTIG K.W. *et al.*, "Probabilistic models for defect initiated fracture in ceramics", *Journal of Nondestructive Evaluation*, vol. 1, no. 2, pp. 111–122, 1980.

[EVA 82] EVANS A.G., "Structural reliability: a processing – dependent phenomenon", *Journal of the American Ceramic Society*, vol. 65, no. 3, pp. 127–137, 1982.

[FER 86] FERBER M.K., TENNERY V.J., WATERS S.B. *et al.*, "Fracture strength characterization of tubular ceramic materials using a simple C-ring geometry", *Journal of Materials Science*, vol. 2, pp. 2628–2632, 1986.

[FOR 12] FORAY G., DESCAMPS-MANDINE A., R'MILI M. *et al.*, "Statistical distributions for glass fibers: correlation between bundle test and AFM-derived flaw size density functions", *Acta Maerialia,* vol. 60, pp. 3711–3718, 2012.

[FRE 68] FREUDENTHAL A.M., "Statistical approach to brittle fracture", in LIEBOVITZ H. (ed.), *Fracture*, Academic Press, New York, 1968.

[FRE 89] FREIMAN S.W., POHANKA R.C., "Review of mechanically related failures of ceramic capacitors and capacitor materials", *Journal of the American Ceramic Society*, vol. 72, pp. 2258–2263, 1989.

[FUK 84] FUKUMORI K., KURAUCHI T., "Static fatigue of rubbery polymers: statistical approach to the failure process", *Journal of Materials Science*, vol. 19, pp. 2501–2512, 1984.

[GOD 86] GODA K., FUKUNAGA H., "The evaluation of the strength distribution of silicon carbide and alumina fibers by a multi-modal Weibull distribution", *Journal of Materials Science*, vol. 21, pp. 4475–4480, 1986.

[GON 00] GONG J., "A new probability index for estimating Weibull modulus for ceramics with least square method", *Journal of Materials Science Letters*, vol. 19, pp. 827–829, 2000.

[GRI 03] GRIGGS J.A., YUNLONG Z., "Determining the confidence intervals of Weibull parameters estimated using a more precise probability estimator", *Journal of Materials Science Letters*, vol. 22, pp. 1771–1773, 2003.

[GUI 96] GUILLAUMAT L., LAMON J., "Fracture statistics applied to modelling the non-linear stress-strain behavior in microcomposites: influence of interfacial parameters", *International Journal of Fracture*, vol. 82, pp. 297–316, 1996.

[GUI 96] GUILLAUMAT L., LAMON J., "Probabilistic-statistical simulation of the nonlinear mechanical behavior of a woven SiC/SiC composite", *Composite Science and Technology*, vol. 56, pp. 803–808, 1996.

[GUM 68] GUMBEL E., *Statistics of Extremes*, Columbia University Press, New York, 1968.

[HEG 91] HEGER A., BRÜCKNER-FOIT A., MUNZ D., "STAU – a computer code to calculate the failure probability of multiaxially loaded ceramic components", *2nd European Ceramic Society Conference*, Euro-Ceramics, 1991.

[HEM 86] HEMPEL H., WIEST H., "Structural analysis and life prediction for ceramic gas turbine components for the Mercedes-Benz research car 2000", *ASME Paper 86-GT-199, Presented at the International Gas Turbine Conference and Exhibit,* Düsseldorf, West Germany, 8–12 June 1986.

[HEN 89] HENSTENBURG R.B., PHOENIX S.L., "Interfacial shear strength studies using the single-filament-composite test. Part II: a probability model and Monte Carlo simulation", *Polymer Composites*, vol. 10, no. 6, pp. 389–408, 1989.

[HIL 92] HILD F., MARQUIS D., "A statistical approach to the rupture of brittle materials", *European Journal of Mechanics, A/Solids*, vol. 11, no. 6, pp. 753–765, 1992.

[HU 85] HU X.-Z., COTTERELL B., MAI Y.-W., "A statistical theory of fracture in a two-phase brittle material", *Proceedings of the Royal Society London*, vol. 401, pp. 251–65, 1985.

[HUI 95] HUI C.-Y., PHOENIX S.L., IBNABDELJALIL M. *et al.*, "An exact closed form solution for fragmentation of Weibull fibers in a single filament composite with applications to fiber-reinforced ceramics", *Journal of the Mechanics and Physics of Solids*, vol. 43, no. 10, pp. 1551–1585, 1995.

[ING 13] INGLIS C.E., "Stresses in the plates due to the presence of cracks and sharp corners", *Transactions of the Institute of Naval Architects*, vol. 55, pp. 219–241, 1913.

[ISO 03] INTERNATIONAL STANDARD ISO/DIS 20501, Fine ceramics (advanced ceramics, advanced technical ceramics) – Weibull statistics of strength data, pp. 12–01, 2003.

[JEU 91] JEULIN D., Modèles morphologiques de structures aléatoires et de changement d'échelles (Morphological models of random structures and of scaling), Thesis, University of Caen, France, 25 April 1991.

[JOH 83] JOHNSON C.A., "Fracture statistics of multiple flaw distributions", *Fracture Mechanics of Ceramics*, vol. 5, pp. 365–386, 1983.

[KAW 86] KAWAMOTO M., SCHIMIZU T., SUZUKI M. *et al.*, "Strength analysis of Si_3N_4 swirl chamber for high power turbocharged diesel engines", in BUNK W., HAUSNER J. (eds), *Proceedings of the 2nd International Symposium on Ceramic Materials and Components for Engines*, pp. 1035–1042, 1986.

[KHA 91] KHALILI A., KROMP K., "Statistical properties of Weibull estimators", *Journal of Materials Science*, vol. 26, pp. 6471–6752, 1991.

[KHA 93] KHANDELWAL P.K., PROVENZANO N.J., SCHNEIDER W.E., "Lifetime prediction methodology for ceramic components of advanced vehicular heat engines", *Proceedings of the Annual Automative Technology Development Contractors, Coordination Meeting*, pp. 199–211, 1993.

[KIR 67] KIRSTEIN A.F., WOOLLEY R.M., "Symmetrical bending of thin circular elastic plates on equally spared point supports", *Journal of Research of the National Bureau of Standards – C. Engineering and Instrumentation*, vol. 71C, no. 1, pp. 1–10, January–March 1967.

[KUS 91] KUSSMAUL K., LAUF S., TURAN K., "FEM-LIFTAP, a finite element lifetime analysis post processor for ceramic components", *4th International Symposium on Ceramic Materials and Components for Engines*, Göteborg, Sweden, 1991.

[LAM 83] LAMON J., EVANS A.G., "Structural analysis of bending strengths for brittle solids: a multiaxial fracture problem", *Journal of the American Ceramic Society*, vol. 66, no. 3, pp. 177–182, 1983.

[LAM 85a] LAMON J., "Structural reliability of ceramics: thermal stress fracture", *Revue Internationale des Hautes Températures et Réfractaires*, vol. 22, pp. 115–127, 1985.

[LAM 85b] LAMON J., Statistical analysis of fracture of silicon nitride using the short span bending technique, ASME Gas Turbine Conference and Exhibit, Houston, Paper no. ASME-85-GT-151, 1985.

[LAM 86] LAMON J., THOREL A., BROUSSAUD D, "Influence of long-term ageing upon the mechanical properties of partially stabilized zirconia (Mg-PSZ) for heat-engine applications", *Journal of Materials Science*, vol. 21, no. 7, pp. 2277–2282, 1986.

[LAM 88a] LAMON J., "Ceramics reliability: statistical analysis of multiaxial failure using the Weibull approach and the multiaxial elemental strength model", *ASME Gas Turbine Conference and Exhibit,* Amsterdam, The Netherlands, Paper N° ASME-88-GT-147, 1988.

[LAM 88b] LAMON J., "Statistical approaches to failure for ceramic reliability assessment", *Journal of the American Ceramic Society*, vol. 71, no. 2, pp. 106–112, 1988.

[LAM 89a] LAMON J., PHERSON D., DOTTA P., 2D and 3D ceramic reliability analysis using CERAM, statistical post processor software technical documentation, Ceramic Design Final Report, Battelle-Geneva, May 1989.

[LAM 89b] LAMON J., MELET J., 2D and 3D ceramic reliability analysis using CERAM, experimental evaluation of reliability analysis by CERAM2D and CERAM3D, Ceramic Design Final Report, Battelle-Geneva, May 1989.

[LAM 90a] LAMON J., "Ceramics reliability: statistical analysis of multiaxial failure using the Weibull approach and the multiaxial elemental strength model", *Journal of the American Ceramic Society*, vol. 73, no. 8, pp. 2204–2212, 1990.

[LAM 90b] LAMON J., Reliability analysis of ceramics using the CERAM computer program, ASME Gas Turbine and Aeroengine Congress, Brussels, Belgium, Paper no. ASME-90-GT-98, June 1990.

[LAM 90c] LAMON J., BENTZEN J.B. BILDE-SORENSEN N. *et al.*, "Structural reliability of ceramics", in BENTZEN J.J. *et al.* (ed.), *Proceedings of the 11th Riso International Symposium on Metallurgy and Materials Science: Structural Ceramics – Processing, Microstructure and Properties*, Riso National Laboratory, Roskilde, Denmark, pp. 39–56, 1990.

[LAM 91] LAMON J., PHERSON D., "Thermal stress failure of ceramics under repeated rapid heatings", *J. Am. Ceram. Soc.*, vol. 74, no. 56, pp. 1188–1196, 1991.

[LAM 93] LAMON J., "Thermal shock behavior of ceramics: probabilistic predictions of failure and damage", in SCHNEIDER G.A., PETZOW G. (eds), *Thermal Shock and Thermal Fatigue Behavior of Advanced Ceramics*, Kluwer Academic Publishers, Holland, pp. 459–471, 1993.

[LAM 94] LAMON J., "Probabilistic failure predictions in ceramics and ceramic matrix fiber reinforced composites", in BRINKMAN CP.R., DUFFY S.F. (eds), *Life Predictions Methodologies and Data for Ceramic Materials*, ASTM STP 1201, American Society for Testing and Materials, Philadelphia, pp. 265–279, 1994.

[LAM 98] LAMON J., THOMMERET B., ERCEVAULT C.P., "Probabilistic-statistical approach to the matrix damage and stress-strain behavior of 2-D woven SiC/SiC ceramic matrix composites (CMCs)", *Journal of the European Ceramic Society*, vol. 18, pp. 1797–1808, 1998.

[LAM 01] LAMON J., "Micromechanics-based approach to the mechanical behavior of brittle matrix composites", *Composites Science and Technology*, vol. 61, pp. 2259–2272, 2001.

[LAM 09] LAMON J., "Stochastic approach to multiple cracking in composite systems based on the extreme value theory", *Composites Science and Technology*, vol. 69, pp. 1607–1614, 2009.

[LAM 10] LAMON J., "Stochastic models of fragmentation of brittle fibers or matrix in composites", *Composites Science and Technology*, vol. 70, pp. 743–751, 2010.

[LAW 82] LAWLESS J.L., *Statistical Models and Methods for Lifetime Data*, John Wiley and Sons, 1982.

[LEW 76] LEWIS III D., OYLER S.M., "An experimental test of Weibull scaling theory", *Journal of the American Ceramic Society*, vol. 59, nos. 11–12, pp. 507–510, 1976.

[LIN 83] LINDGREN L.C., HEITMAN P.W., THRASHER S.R., Ceramic application in gas turbine engines, SAE Technical Paper Series, 831520, 1983.

[LIS 97a] LISSART N., LAMON J., "Statistical analysis of failure of SiC fibers in the presence of bimodal flaw populations", *Journal of Materials Science*, vol. 32, pp. 6107–6117, 1997.

[LIS 97b] LISSART N., LAMON J., "Damage and failure in ceramic matrix minicomposites: experimental study and model", *Acta Mater.*, vol. 45, no. 3, pp. 1025–1044, 1997.

[LU 02] LU C., DANZER R., FISCHER F.D., "Fracture statistics of brittle materials: Weibull or normal distribution", *Physical Review E*, vol. 65, p. 06702, 2002.

[MAN 55] MANSON S.S., SMITH R.W., "Theory of thermal shock resistance of brittle materials based on Weibull's statistical theory of strength", *Journal of the American Ceramic Society*, vol. 38, pp. 18–27, 1955.

[MAT 76] MATTHEWS J.R., MCCLINTOCK F.A., SHACK W.J., "Statistical determination of surface flaw density in brittle materials", *Journal of the American Ceramic Society*, vol. 59, no. 7–8, pp. 304–308, 1976.

[MCC 66] MCCLINTOCK F.A., ARGON A.S., *Mechanical Behavior of Materials*, Addison-Wesley Publishing Company Inc., Reading, MA, 1966.

[NEL 82] NELSON W., *Applied Life Data Analysis*, John Wiley & Sons, New York, 1982.

[NEM 90] NEMETH N.N., MANDERSCHEID J.M., GYEKENYESI J.P., Ceramics analysis and reliability evaluation of structures (CARES) users and programmers manual, NASA TP-2916, 1990.

[OCH 95] OCHAI S., HOJO M., "Multiple cracking of a coating layer and its influence on fiber strength", *Journal of Materials Science*, vol. 30, pp. 09–513, 1995.

[OKA 08] OKABE T., NISHIKAWA N., CURTIN W.A., "Estimation of statistical strength distribution of Carborundum polycrystalline Si C fiber using the single fiber composite with consideration of the matrix hardening", *Composites Science and Technology*, vol. 68, pp. 3067–3072, 2008.

[OKA 99] OKABE T., TAKEDA N., KOMOTORI J. *et al.*, "A new fracture mechanics model for multiple matrix cracks of SiC reinforced brittle matrix composites", *Acta Materialia*, vol. 47, pp. 4299–4309, 1999.

[PAI 05] PAILLER F., LAMON J., "Micromechanics-based model of fatigue/oxidation for ceramic matrix composites", *Composites Science and Technology*, vol. 65, pp. 369–374, 2005.

[PEI 26] PEIRCE F.T., "Tensile tests for cotton yarns – the weakest link – theorems on the strength of long and of composite specimens", *Journal of the Textile Institute*, Transactions, vol. 17, no. 7, pp. 355–368, 1926.

[PET 87] PETROVIC J.J., "Weibull statistical fracture theory for the fracture of ceramics", *Metallurgical Transactions A*, vol. 18A, pp. 1829–1834, 1987.

[PIN 92] PINEAU A., FRANÇOIS D., ZAOUI A., *Comportement Mécanique des Matériaux*, Hermes, Paris, 1992.

[POL 71] POLONIECKI J.D., WILSHAW T.R., "Determination of surface crack size densities in glass", *Nature*, vol. 229, p. 226, 1971.

[PRA 99] PRABHAKARAN R., SAHA M., "Tension, compression and flexure properties of pultruded composites", *Proceedings of the 12th International Conference on Composite Materials*, Paris, 1999.

[QUI 05] QUINN G., IVES L.K., JAHANMIR S., *Machining Cracks in Finished Ceramics, Key Engineering Materials*, Trans. Tech Publications, Switzerland, vol. 290, pp. 1–13, 2005.

[RIC 02] RICE R.W., "Monolithic and composite ceramic machining flaw-microstructure-strength effects: model evaluation", *Journal of the European Ceramic Society*, vol. 22, nos. 9–10, pp. 1411–1424, 2002.

[RIC 79] RICE R.W., HOCKEY B.J. (eds), The Science of Ceramic Machining and Surface Finishing II, National Bureau of Standards, October 1979.

[RMI 12] R'MILI M., GODIN N., LAMON J., "Flaw strength distributions and statistical parameters for ceramic fibers: the normal distribution", *Physical Review E*, vol. 85, no. 5, pp. 1106–1112, 2012.

[ROU 09] ROUSSET G., LAMON J., MARTIN E., "In situ fiber strength determination in metal matrix composites", *Composites Science and Technology*, vol. 69, pp. 2580–2586, 2009.

[RUF 84] RUFIN A., SAMOS D., BOLLARD R., "Statistical prediction models for brittle materials", *AIAA Journal*, vol. 22, no. 0.1, pp. 135–140, 1984.

[SAW 87] SAWYER L.L., JAMIESON M., BRIKOWSKI D. *et al.*, "Strength, structure and fracture properties of ceramic fibers produced from polymeric precursors: I: base-line studies", *Journal of the American Ceramic Society*, vol. 70, pp. 789–810, 1987.

[SCH 96] SCHENK B., Experimental and numerical methods for determining the lifetime of ceramic rotor components of small high temperature gas turbines (en allemand), Wissenschaft & Technik Verlag, Berlin, Germany, 1996.

[SCH 99] SCHENK B., BREHM P., MENON M.N. *et al.*, "Status of the CERAMIC/ERICA probabilistic life prediction codes development for structural ceramic applications", *ASME Paper 99-GT-318, Presented at the International Gas Turbine and Aeroengine Congress and Exposition*, Indianapolis, 7–10 June 1999.

[SHE 80] SHETTY D.K., ROSENFIELD A.R., MCGUIRE P. *et al.*, "Baxial flexure tests for ceramics", *Ceramic Bulletin*, vol. 52, no. 2, pp. 1193–1197, 1980.

[SHE 83] SHETTY D.K., ROSENFIELD A.R., DUCKWORTH W.H. *et al.*, "A biaxial-flexure test for evaluating ceramic strengths", *Journal of the American Ceramic Society*, vol. 66, no. 1, pp. 36–42, 1983.

[SHE 84] SHETTY D.K., ROSENFIELD A.R., DUCKWORTH W.H., Statistical analysis of size and stress state effects on the strength of an alumina ceramic, Special Technical Testing Publication 844, American Society for Testing and Materials, Philadelphia, 1984.

[SMA 90] SMART J., "The determination of failure probability using Weibull probability statistics and the finite element method", *Res Mechanica*, vol. 31, pp. 205–219, 1990.

[SNE 89] SNEDDEN J.D., SINCLAIR C.D., "Statistical mapping and analysis of engineering ceramics data", in COCKS A.C.F., PONTER A.R.S. (eds), *Mechanics of Creep Brittle Materials 1*, Elsevier Applied Science Publishers, London, 1989.

[STU 91] STÜRMER G., SCHULZ A., WITTIG S., "Lifetime prediction for ceramic gas turbine components", *ASME Paper 91-GT-96, Presented at the 36th International Gas Turbine and Aeroengine Congress and Exposition*, Orlando, FL, 1991.

[SUL 86] SULLIVAN J.D., LAUZON P.H., "Experimental probability estimators for Weibull plots", *Journal of Materials Science Letters*, vol. 5, pp. 1245–1247, 1986.

[TIM 59] TIMOSHENKO S., WOINOWSKY-KRIEGER S., *Theory of Plates and Shells*, 2nd ed., McGraw Hill, New York, 1959.

[TIM 70] TIMOSHENKO S.P., GOODIER J.N., *Theory of Elasticity*, McGraw-Hill, New York, 1970.

[TRU 79] TRUSTRUM K., DE JAYATILAKA S.A., "On estimating the Weibull modulus for a brittle material", *Journal of Materials Science*, vol. 14, pp. 1080–1084, 1979.

[VIT 63] VITMAN F.F., PUKH V.P., "A method for determining the strength of sheet glass", *Zavod Lab.*, vol. 29, no. 7, pp. 863–867, 1963.

[WAG 89] WAGNER H.D., "Stochastic concepts in the study of size effects in the mechanical strength of highly oriented polymeric materials", *Journal of Polymer Science, Part B: Polymer Physics*, vol. 27, pp. 115–149, 1989.

[WAG 90] WAGNER H.D., EITAN A., "Interpretation of the fragmentation phenomenon in single-filament composites experiments", *Applied Physics Letters*, vol. 56, 1990.

[WEI 39] WEIBULL W., A statistical theory of the strength of materials, Ingeniorsvetenkapsakodemiens hadlinger NR 151, Generalstabens Litografiska Anstaltz Förlag, Stockholm, 1939.

[WEI 51] WEIBULL W., "A statistical distribution function of wide applicability", *Journal of Applied Mechanics*, vol. 18, pp. 293–297, 1951.

[YAM 84] YAMASHITA K., KOUMOTO K., YANAGIDA H. *et al.*, "Analogy between mechanical and dielectric strength distributions for $BaTiO_3$ ceramics", *Journal of the American Ceramic Society*, vol. 67, no. 2, pp. C31–C33, 1984.

Index

S, T, W